PLAN IT, DIG IT, BUILD IT!

Your Step-By-Step Guide to Landscape Projects

David Sauter

THOMSON

DELMAR LEARNING

Australia Canada Mexico Singapore Spain United Kingdom United States

THOMSON

DELMAR LEARNING

Plan It, Dig It, Build It!
Your Step-By-Step Guide to Landscape Projects
David Sauter

Business Unit Executive Director:
Susan L. Simpfenderfer

Executive Production Manager:
Wendy A. Troeger

Executive Marketing Manager:
Donna J. Lewis

Acquisitions Editor:
Zina M. Lawrence

Production Manager:
Carolyn Miller

Channel Manager:
Nigar Hale

Editorial Assistant:
Rebecca Switts

Production Editor:
Kathryn B. Kucharek

For permission to use material from this text or product, contact us by
Tel   (800) 730-2214
Fax   (800) 730-2215
http://www.thomsonrights.com

Library of Congress Cataloging-in-Publication Data

Sauter, David.
    Plan it, dig it, build it! : your step-by-step guide to landscape projects / David Sauter.
        p. c.m.
    Includes index.
    ISBN 1-4018-1044-6
    1. Garden structures—Amateurs' manuals.
2. Landscape constructions—Amateurs' manuals. I. Title.

TH4961 .S28 2002
690.89—dc21   2002035042

NOTICE TO THE READER

Publisher does not warrant or guarantee any of the products described herein or perform any independent analysis in connection with any of the product information contained herein. Publisher does not assume, and expressly disclaims, any obligation to obtain and include information other than that provided to it by the manufacturer.

The reader is expressly warned to consider and adopt all safety precautions that might be indicated by the activities herein and to avoid all potential hazards. By following the instructions contained herein, the reader willingly assumes all risks in connection with such instructions.

The Publisher makes no representation or warranties of any kind, including but not limited to, the warranties of fitness for particular purpose or merchantability, nor are any such representations implied with respect to the material set forth herein, and the publisher takes no responsibility with respect to such material. The Publisher shall not be liable for any special, consequential, or exemplary damages resulting, in whole or part, from the readers' use of, or reliance upon, this material.

Contents

Chapter 3
PREPARING THE SITE FOR CONSTRUCTION   48

Chapter 4
GRADING, EROSION CONTROL, AND DRAINAGE 62

Chapter 5
LANDSCAPE RETAINING WALLS 89

Chapter 8
FENCES AND FREE-STANDING WALLS   **189**

Chapter 9
WATER, EDGING, LIGHTING, AND OTHER SITE AMENITIES

229

Chapter 10
ADVANCED PROJECTS 269

APPENDIXES 273

GLOSSARY 284
INDEX 291

Gardening and landscaping are two fields in which homeowners often take an active role in the improvement of their property.
Some participate for the esthetic, and occasionally, financial rewards
that come from an attractive landscape. Some landscape to address the
functional needs of their site. Many participate out of the sheer joy of
completing projects and watching their creations grow. The effort
involved in landscaping is seldom considered "work" by those who love
gardening and building.

Landscape projects are filled with benefits beyond love and money for
both the individual and the family. The physical activity required to
build projects helps keep a person physically toned. The problem
solving required to put a project together keeps you mentally alert.
Building and growing plants have psychological benefits. Working with
one's family to complete projects creates opportunities to bond.
Whether you have a love for improving your environment, or enjoy the
therapeutic benefits of landscaping, at some point in time almost
everyone has a desire to enhance his or her lawn, garden, or yard.

Despite their interest in landscaping, many homeowners nevertheless
avoid many projects because they are not sure of their ability to complete them properly. They lack the confidence that comes from experience and guidance. This book is intended to provide the bridge
between wishful thinking and tangible results. Landscaping is not
impossible. Learning how to build hardscape such as terraces, walls,
patios, or pools begins with a good plan and some basic instructions.
Remember that everyone had to start with a first project. Once you've
accomplished your first, you'll feel the confidence to move on to

another! This book is designed to help those of you who have a desire to shape and cultivate your surroundings do just that.

If your barrier is a design issue, such as not knowing what to put where, rest assured that you are the best judge of how you will use your space. You know what activities your family enjoys in your yard. You know where the sunny spots and quiet niches are located. The process of design requires selecting and placing outdoor use areas, pathways, and plantings. Review the ideas in this book and observe how you and your family use your site. Experiment with the placement of patios, fences, walks, and other landscape elements to see whether they fit with how you use your site. If the challenge of design is still overwhelming or you need confirmation of your ideas, hire a design consultant for a couple hours to tell you whether your project is adequate and the location appropriate. After possibly modifying your plans based on the design consultant's advice, you can proceed with your project, knowing that your plans will work as you envision them working and that you only need to work on the details.

THE PROCESS OF LANDSCAPE CONSTRUCTION

The activities included in this book are designed to provide you with a number of do-it-yourself projects that can enhance your home landscape. Possible projects range from simply setting a bench in the front yard to adding a paved patio in your favorite area of the yard. Like any home project, the quality of the work will benefit from an orderly approach to the activity.

One aspect of successful landscaping is following a logical progression through each phase of the landscape construction process. The chapters and projects in this text are ordered according to the way a landscape project typically is constructed, beginning with planning and proceeding through to the finishing touches. When one is excited about a project it is tempting to begin without engaging in any planning, and the results can be disastrous. Therefore, Chapters 1 and 2 will acquaint you with the safety and planning issues you should consider before undertaking a project. Most projects require removing some unwanted existing landscaping and manipulating the grade of the site to accommodate improvements. Chapters 3 and 4 will advise you about preparing your site for construction once planning is complete. Work done under the guidance of these chapters will help ensure that your projects end up actually improving your site.

Chapters 5 through 8 cover the main elements of hardscape development. These chapters will take you step by step through building a retaining wall, paving an outdoor area, building with wood, or creating a fence or wall for your property. Concentrating on specific projects, these chapters will help you implement ideas that will fit almost any site. Chapter 9 addresses specialized landscape elements as well as finishing touches and accents that can bring life to many landscapes: the water features, lighting, edging, steps, and other amenities described in this book will enhance your completed landscape project.

Words set in bold type throughout the book are defined in the glossary.

Included in the text are over 40 sample projects that are suitable for most homeowners. With some experience in operating tools and some guidance in implementing the project, you can make these landscaping ideas a reality.

THE PROJECT FORMAT

The planning and implementing of each idea in this book is presented in a "project" format. This format presents each project as a step-by-step installation, complete with a list of the materials and tools you will need to complete the work. These projects are based on a standard design that will fit most landscape settings but can be customized by the owner by modifying the list of tools and materials needed.

To help you decide which projects are suitable for your skill level, the difficulty of each "project" is rated as easy, moderate, or challenging. An "easy" rating indicates that the project will require only basic experience in the operation of the tools necessary to complete. A "moderate" rating indicates that you should be familiar with the operation of any special tools required for the project, or that it requires moderate physical activity, such as digging or lifting. A project rated as "challenging" may require you to operate specialty tools, deal with complex instructions, or engage in extensive physical activity. It may also require assistance from another person to complete because of the physical activity involved.

To enhance the basic project, Chapter 10 suggests ways to "spice up" your landscape-project selections. Different designs or additional details will show how the basic project can evolve into something more comprehensive or artistic. Use these additions as guidelines for your own creativity in improving upon our ideas.

WHEN A PROFESSIONAL SHOULD BE CONSULTED

Most homeowners have had the feeling at one time or another that no project is above their skills. Although most of the projects in this book are appropriate to the skill level of most homeowners, there may be circumstances when you will need help. Some situations in which a professional should be involved include custom designs, installing utility lines, sizing structural members of wood construction, and extensive overhead or trenching work. Some tasks, such as pouring concrete, building decks and gazebos, and operating certain pieces of equipment, have been omitted from this text. Although these are important elements of many landscaping projects, the risks inherent in them should be left to the qualified practitioner.

To avoid the risk of injury or worse, it is important to understand which aspects of a project are within your skill level and when a professional skilled in that work should be contracted. The book frequently recommends that certain tasks should be completed by qualified professionals. In addition to these recommendations, it presents Cautions when a task involves serious dangers.

Cautions Regarding the Use of this Book

In addition to seeking the assistance of a professional, following are a list of general cautions to review before beginning projects outlined in this text:

- Even the simplest construction activity can present safety and health hazards if proper precautions are not observed. Chapter 1, as well as Cautions placed throughout the book, alert you to some of the potential hazards of construction activities. The reader, however, assumes all risk for activities that they undertake.

- Always wear proper safety equipment and clothing when working on a project.

- Regulations regarding projects will vary from region to region. You must verify with local building professionals (municipal, county, and other authorities with jurisdiction over your project site) the appropriateness of recommendations made in this text. Particular areas to review with these officials include the type, dimensions, and location of the project on a property. See Chapter 2, Planning Your Project, for more information.

- This book assumes that you possess the basic skills necessary to safely perform construction activities and operate construction equipment. If you are not familiar with construction techniques, obtain instruction or assistance in these areas before attempting to undertake the activities described in this book.

- Manufacturers' instructions or manuals supersede any recommendations made in this text. Those instructions and manuals should be thoroughly read and understood before using any tool, product, or procedure.

- All data and information contained in this text are believed to be reliable; however, no warranty of any kind, express or implied, is made with respect to the data, analysis, information, or applications contained herein; and the use of any such data, analysis, information or applications is at the user's sole risk and expense. Neither I nor the publisher of this book assumes any liability and expressly disclaims liability, including, without limitation, liability for negligence resulting from the use of the data, analysis, information, or applications contained in this text.

GETTING STARTED

After you have determined which projects you want to undertake and have thoroughly read and understood Chapters 1 and 2, proceed through the text in order or jump to the section that discusses the project you are interested in. When you feel confident about your skills, try some of the alternative projects or comprehensive projects suggested at the end of this book. Reaping the many benefits of landscaping requires taking that first step, and with this text you have no reason to postpone that journey any longer. Pick your project, plan your work, line up the tools and materials, and get started!

Chapter 1

PLAN IT, DIG IT, BUILD IT!: SAFELY

Most of us know of someone who has hurt himself or herself during a construction project, or whose back hurts from years of bending and lifting. Having spent years working in the landscaping and construction field, I have many acquaintances who are limited in their activities because of the rigors of landscape work. Some examples are the gardener who has improperly lifted too many petunia jumbo packs, the contractor with saw-cut scars on his hands, and some unfortunate ones who are now in different jobs as a result of accidents and injuries sustained while doing landscape work. Sadly, many of these acquaintances could have avoided their present situation had they properly prepared themselves to safely perform their work.

When properly performed, most activities required to complete a landscape project pose limited risk. But, like any construction activity, landscaping also carries with it the risk of serious injury, or worse, if you are unaware of its hazards. In the introduction to this book, the therapeutic value of performing landscape tasks was discussed. However, the many benefits and rewards of landscaping can be quickly offset by engaging in activities that lead to a serious or debilitating injury.

Fortunately, many of the hazards of the landscape workplace are identifiable, and, if proper procedures are followed, avoidable. Other activities, however, hold hidden dangers that may not be apparent unless you have experienced them. Because almost every potential hazard of a construction site can be identified, you should develop an approach to working that forces you to assess each planned activity for potential hazards before beginning it. Failure to think through a project may actually *create* a hazard in an activity that was previously considered safe.

This chapter outlines several common landscaping hazards, along with methods for reducing the risks associated with them. Although not every hazard is defined, working safely can assure that the therapy you receive is from completing the project rather than from rehabilitation necessitated by an injury.

ACCIDENT PREVENTION

The prevention of accidents begins with recognizing hazards in the workplace. Although this exercise could overwhelm a person thinking of every possible way they could place himself or herself at risk, you should examine a few common areas in which prevention is the best approach.

Tool And Equipment Operation

For a safe beginning, ensure that you are properly trained and have experience in the use of the tools and equipment required for a project before engaging in any construction activities. Landscape work uses many common household tools that you have probably used before, such as hammers, saws, drills, screwdrivers; but using equipment such as **chainsaws**, **sod cutters**, **post-hole augers**, and specialized landscaping tools may require that you obtain special training in their operation before you begin.

You should also be familiar with the proper techniques for doing construction work, including sawing, hammering, digging, and the like. To develop basic skills in the areas mentioned above, you might explore courses offered by a local community-college program in horticulture or building trades, or attend seminars sponsored by building- supply companies. Programs that offer hands-on opportunities to learn provide the best instruction.

CAUTION

■ This book is intended to provide project guidelines to those who already have experience in the use of the tools and equipment required for landscape projects. It is not intended as a training manual in basic construction techniques and tool operation. If you are not familiar with operating this equipment, seek training before attempting to build the projects described in this book.

- If you engage in projects without knowing how to safely use the tools and equipment required, or without knowledge of proper construction techniques, you are putting your health and safety at risk.

Reviewing Safety Information

Before beginning a project, review all instructions and warnings provided by equipment manufacturers and product suppliers. This information is intended to protect you from injury, particularly from hazards that might not be obvious. The instructions for virtually every tool, material, and product that you will use in landscaping provide some information regarding its safe use.

First Aid

I recommend that anyone who engages in landscaping activities also consider getting basic first aid and CPR training. Knowing how to treat simple pinches and cuts should be second nature if you do many home projects. It is also important to know how to determine when emergency help should be summoned. If you are not familiar with health hazards such as allergic reactions to stings or how to recognize heat stroke, a first aid class is also essential. You should keep a first aid kit near the work area and have a handy means to communicate emergencies if necessary. Precious time can be lost in an emergency searching for a phone.

Remember that chronic health problems may be caused or aggravated by working conditions or activities; for example, dust or pollen from work sites may trigger allergies and asthma attacks. I recommend that you familiarize yourself with the appropriate response for such conditions, and be sure that your tetanus booster shot is current before beginning your work.

PERSONAL PROTECTION

The most important tool you have is your body. Protecting your body from injury is necessary to ensure that the projects you build can be enjoyed for a lifetime. Following are some issues you may face when performing landscape work.

Use of Proper Clothing and Safety Equipment

Protecting your body is a logical starting place when attempting to reduce the possibility of injury. Essential equipment for building landscape projects includes proper work boots with steel toe protection, heavy-duty work gloves, and eye protection. Also consider using a hard hat whenever work is overhead; appropriate back support for lifting activities, and skin protection when you expect to be exposed to the sun for extended periods. Dust masks and ear plugs are also important when performing landscape construction tasks that involve constant noise and/or dust. Kneepads will prove helpful when performing tasks that require long periods of kneeling.

Heat and Cold Injury

Common sense suggests that you dress for the conditions outside, so consider appropriate dress when working on landscape projects for long periods in the sun or cold. Because you may not notice slight increases in your body temperature, threats from heat injury should be taken seriously. In addition to protecting your skin from burn and excess exposure, be sure to protect yourself from heat exhaustion and heat stroke when activities are planned for very hot days. Plan appropriate work hours, ample breaks, and make sure plenty of water is available when temperatures climb.

Wildlife

Working in the outdoor environment will bring you in constant contact with wildlife. Most wildlife pose limited threats, but when we disturb their habitats, their attempts to defend themselves can be dangerous. Common problems from wildlife include stings or bites from bees, wasps, spiders, rodents, snakes, and scorpions, as well as possible attacks from larger animals, depending on your locale. When performing site-preparation tasks that require the disturbance or removal of potential animal habitats, proceed with caution. Old structures, brush piles, dead trees, shrubs, or drainage ways may be unsightly nuisances to a homeowner but a wonderful home for animals. To locate signs of animal habitation, survey the site carefully for tracks, nests, webs, or feces before beginning work. Carry insect spray to ward off any attacks that may be provoked. Examine equipment, tools, and clothing left at the site to ensure that wildlife has not adopted the items for housing while it was left unattended.

If you discover wildlife habitats, whether you consider the animals that inhabit them threatening or not, consider contacting local animal shelters or protection centers. Some organizations are willing to relocate animals to new habitats before problems arise.

Hazardous Plants

Be aware of any plants that may trigger itching, blistering, or allergic reactions if they come into contact with your skin. Obtain positive identification of all suspected plants and take appropriate precautions. Assistance in identification of plants (and in many cases, wildlife) may be available from your county **agricultural extension** offices. Remove or avoid poisonous plants, or wear proper protective clothing in order to avoid such direct contact. If burning caustic plants, do not expose yourself to the smoke.

Back Injury

Many of the actions and motions required for landscaping work involve inherent risk. The improper lifting of heavy objects can produce strain on muscle groups that were not intended to lift heavy weights and can create an immediate—and serious— injury. Your projects may require you to frequently pick up materials and tools. After improperly performing such a motion numerous times without injury, you may one day suffer an injury from lifting a light object. Repetitive improper lifting of light objects, even with muscle groups that consistently lift such weights, creates strain that can accumulate and lead to injury.

Learning how to lift properly and following some simple rules will help you reduce the risk of back injury resulting from landscape construction activities. Also, performing some simple stretching exercises before beginning work and following long periods of repetitive motion can help reduce problems (Figure 1-1).

Proper lifting techniques are designed to direct the strain of lifting to the proper muscle groups. The following techniques may help with your common lifting tasks:

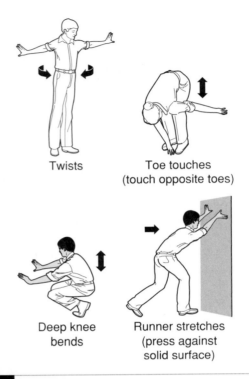

Twists

Toe touches
(touch opposite toes)

Deep knee
bends

Runner stretches
(press against
solid surface)

Figure 1-1   *Warm-up stretching exercises.*

1. Lift by bending at the knees rather than bending at the waist. This transfers the stress to the thigh muscles, which are bigger, stronger, and better adapted to lifting such weight (Figure 1-2).

2. Avoid twisting motions while lifting. Move the feet rather than twisting.

3. When lifting, hold the load close to your body. Holding the load at arms length greatly increases the stress on the back.

4. Get help when lifting heavy objects.

5. Wear back-support braces approved by OSHA or other appropriate health organizations.

6. Properly warm up and rest when engaging in repetitive motions, such as brick laying or planting.

Extended sitting and kneeling can also cause muscle or joint injury. When you sit or kneel for longer than 15 minutes, take a break to stand and stretch. Arrange to have comfortable seating and cushioning for back and knees when performing tasks that require long periods of sitting or kneeling. Additional suggestions for preventing back injuries are available from physicians, physical therapists, and various injury-prevention and rehabilitation organizations.

Figure 1-2 *Proper lifting by bending at the knees.*

SPECIFIC CONSTRUCTION HAZARDS

In addition to identifying and preventing potential hazards, some activities carry with them predictable risks. When engaging in the activities below be aware of the specific dangers each may pose.

Working Around Utilities

Chapter 2 introduces a method for locating utilities on a site. Because the penalty for disrupting a utility could be injury, death, or financial hardship, take the time to locate all utilities before beginning any construction.

Working Below Grade

Although it is unlikely that a homeowner will engage in a project that will require digging to extreme depths, you should be aware that trenches pose a risk of collapse whenever the soil is unstable or the trench deep. Special precautions should be taken any time you are digging and the excavation is waist deep or deeper. Unstable soils should be dug with side slopes at a 45-degree angle rather than vertically. Sheets of plywood can also be used to shore up, or hold in place, the sides of a shallow excavation. When performing a project that requires a trench, use caution when bending or kneeling down to perform work tasks near the bottom. Always work with a partner when doing this type of work, and ensure that one of the partners is outside the trench while the other is digging.

Operating Electrical Equipment

Exercise care when using electrical equipment in landscape applications. Whenever you have electrical cords strung around a construction site, it is easy to accidentally sever a cord, creating a potential for shock. Water ponded on a site is also a potential hazard involved in using electrical equipment outdoors. Verify that the electrical equipment you are using has **ground fault circuit interrupt (GFCI)** protection against shock and electrocution. The ground-fault protection feature is installed by an **electrical contractor** at an outlet or on an entire circuit and will shut a circuit off if it detects a **short (circuit).** Avoid using electrical equipment in wet weather and on sites that contain standing water.

Operating Power Sprayers

Always exercise caution when using power washers and sprayers. Never aim the spray at exposed skin or body parts. A direct spray at high pressure can cause serious damage to eyes and skin, including contusion (bruising) and cutting of the skin or the penetration of materials into the skin.

Operating Tractors or Loading Equipment

If you use a tractor or **loading equipment (skidsteer)** during your work on a project, understand how to properly operate the equipment before beginning work. Most landscape equipment designed to carry materials can tip if the load is lifted too high or the equipment is operated on a slope. When using such equipment, it is necessary to carry loads close to the ground. When operating on slopes it is preferable to operate in a top-to-bottom direction rather than from side to side.

Clean Worksite

Keeping your worksite clean will reduce accidents. Picking up tools and scrap materials can help to avoid accidental slips and falls. Survey the path of backing vehicles before they enter the site to ensure that no critical or dangerous items are in the way.

Working With Chemicals

When painting, staining, and cleaning a site you can be seriously injured or burned from the misuse of chemicals. Following a few simple rules when working with acids, wood preservatives, stains or paints, and similar chemical agents can reduce the potential risk:

- Always wear proper protective clothing when mixing and applying chemicals. Eye and face protection, rubber gloves, and complete covering of body and limbs will afford protection when working with chemicals.

- Mix and use chemicals in open areas with ample ventilation to avoid being overcome by high concentrations of fumes.

- Follow the procedures recommended by the manufacturer when mixing, storing, and applying any chemicals.

In summary, always plan your work carefully, concentrate fully on the task you are engaged in, and use caution and simple common sense when using tools, equipment, and materials, or when performing actions with which you may be unfamiliar and that may be potentially dangerous.

PLANNING YOUR PROJECT

When working on landscape projects, don't consider the word *plan* a negative term. Undertaking critical thought before picking up the shovel and hammer will help to prevent problems and improve the quality of the project. Not all projects require extensive planning to be successful, but proceeding with any project without adequate preparation increases your risk for disappointment. Even simple brainstorms sketched on the back of a napkin deserve critical scrutiny; ask questions like "Where exactly is that gas line?" and "Is this really my property line?"

For most of your projects, consider preparing a drawing called a **landscape plan** (sometimes termed a site plan) that sketches out the improvements. The landscape plan lets the designer and builder explore ideas, estimate quantities, and show how the work will impact the site. The planning process also involves reviewing regulations, locating utilities, and gaining approval from any building officials who may need to review your project. With work and refinement your landscape plan can guide the installation of an attractive and functional project. If you reject an idea or change it significantly because of issues encountered during this phase, then planning has performed its purpose.

Although it is best to consider getting professional help for complex projects, ample assistance is available for simple, single-purpose projects. Reviewing design magazines or books, searching websites, or consulting manufacturers' recommendations will provide many good ideas for the design of landscape elements. Consider hiring a professional designer when the project includes terraces, walls over three feet tall, extensive paving, solving significant drainage problems, determining sizes for lumber, connecting a project to an existing structure, or working with or near utilities. Professional designers can also help when you find it impossible to generate any ideas on your own. Like so many other professions, buying a few hours of a designer's time may save you a lot of time better spent on other activities.

READING LANDSCAPE PLANS

Projects of all sizes benefit from a drawing that documents the planned work. Large landscape projects may require extensive documentation, but you can often complete small residential projects with a scaled plan. Landscape plans are typically drawn in a form called a **plan view**. A plan view shows a project as if looking straight down at it from above, similar to the view a bird would have when flying over the project. Only two dimensions of site elements, length and width, are visible in a plan view, as shown in Figure 2-1. The height of objects can only be interpreted when measurements or shadows are added.

The key elements of the landscape plan are symbols and labels, which are used to identify the various elements of the project. Symbols typically used for landscape plans include circles to show plants, with the sizes of the circles providing an approximate indication of the plants'

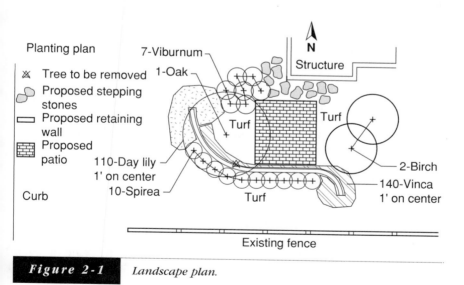

| Figure 2-1 | Landscape plan. |

size. Structures are usually drawn with dark lines so that they stand out, whereas paving is indicated by edge lines and textures (dots, lines, or shading). Fences, walls, stairs, and other landmarks on the site are indicated by symbols and lines showing their location and size. To assist you in reading the plan, many of the symbols just described also include a label.

Reading the plan is the first step to understanding a project, and to accurately draw and interpret plans you must understand the concept of **drawing scale**. Because most drawings cannot be drawn full size, the plan must be reduced to fit on a sheet of paper that is convenient to carry and read. To reduce the size in a manner that will be consistent and measurable for all drawings, a scale is used.

Simply put, scale means that a certain measurement on a drawing is equal to a specified distance on the actual site. An example would be a plan done in 1/4" scale, where 1/4" on a plan is equal to 1' on a site. If a walkway on a plan measured 3/4" wide, the actual width of the walk should be 3'. Scale is useful because any object on a scaled plan can be measured in this manner, and conversely, any object on a site can be drawn to scale.

Scale in landscape drawings is typically expressed as one inch on the plan equals a certain number of feet on the actual site. Two types of scales are typically used for plans: an **architect's scale** using fractional divisions, and an **engineer's scale**, using multiples of ten. Your choice of scale will depend on the size of the object or site being represented. Most landscape projects use a scale of one-eighth inch or one-quarter inch to equal one foot on the actual site (as noted in the example above).

Large projects like an entire yard could be drawn in engineer's scale. Engineer's scales use 1" (on the plan) = 10' (on the actual site), 1" = 20', and similar increments, up to 1" = 60'. Using a scale in which 1" = 10', a walk that measures 1" wide on the plan would be 10' on an actual site. To illustrate both architect's and engineer's scales, lines drawn with both are shown in Figure 2-2.

To quickly measure plans, special instruments, also called scales, are available. Rather than having to constantly measure the number of inches and multiply by a conversion number, scales indicate the number of actual feet according to standard architect's- or engineer's- scale increments. Because landscape drawings can be drawn in either type of scale, whoever draws the plan should indicate the scale on each drawing to assist those who read the plan.

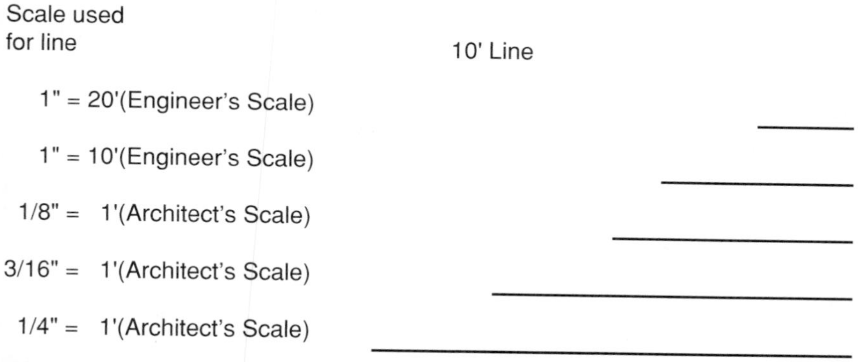

Scale used
for line

10' Line

1" = 20'(Engineer's Scale)

1" = 10'(Engineer's Scale)

1/8" = 1'(Architect's Scale)

3/16" = 1'(Architect's Scale)

1/4" = 1'(Architect's Scale)

Figure 2-2   *10' lines drawn with common architect's and engineer's scales.*

The reading of scales, particularly the architect's scale, takes some practice, but they are invaluable tools for interpreting plans. Scales can be purchased at hobby shops, at stores that sell drafting instruments, and at many college bookstores. If no scales are available, you can interpret drawings prepared in 1/8" and 1/4" scales using a standard ruler. When using a standard ruler, measure lines by identifying and counting the subdivisions (e.g., for 1/4" scale, count each 1/4" as 1'). For long measurements, you will have to count each subdivision to get the complete length.

Reading and measuring with architect's and engineer's scales

Time: 1 hour

Level: Easy (6 steps).

Tools Needed:

1. Architect's scale.

2. Engineer's scale.

continued

3. Drawings prepared using architect's and engineer's scales for practice. Try to find some old blueprints or contact a designer and ask for some old drawings.

4. Paper.

5. Pencil.

No materials required.

Directions:

Measurement using an architect's scale:

1. Identify the scale of the drawing you plan to use.

2. Locate the scale that matches the drawing being interpreted. The fractions for the scales are printed at the ends. (Architect's scales have two scales running in opposite directions along the same edge of the scale. For one scale you read the top set of numbers and for the other scale, the bottom set. The correct set of numbers for your scale should begin with a zero near the fraction indicating your scale.) See Figure 2-3, step A.

3. Identify a line to be measured.

4. Place the scale next to the line being measured so that one end of the line is in the inch measurement area; this is the finely divided scale to the left or right of the zero (Figure 2-3, step B). Inch markings on the architect's scale may represent inches or fractions of inches, depending on the accuracy of the scale. The inch portion of the scale always contains one foot divided into various proportional segments.

5. Adjust the scale left or right until the end of the line being measured rests directly on a foot mark along the scale while the other end remains within the inch measurement area.

6. Read the foot reading and inch reading to obtain the measurement. (Figure 2-3, step C).

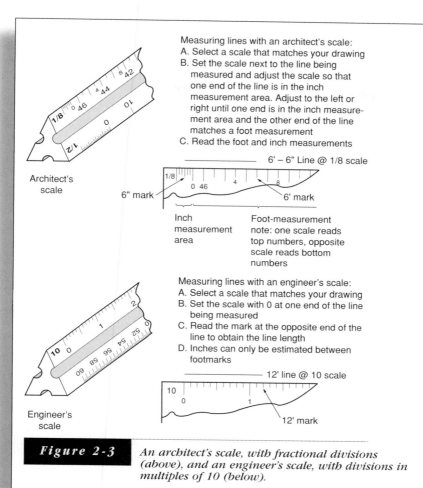

Measuring lines with an architect's scale:
A. Select a scale that matches your drawing
B. Set the scale next to the line being measured and adjust the scale so that one end of the line is in the inch measurement area. Adjust to the left or right until one end is in the inch measurement area and the other end of the line matches a foot measurement
C. Read the foot and inch measurements

6' – 6" Line @ 1/8 scale

Architect's scale

6" mark

6' mark

Inch measurement area

Foot-measurement note: one scale reads top numbers, opposite scale reads bottom numbers

Measuring lines with an engineer's scale:
A. Select a scale that matches your drawing
B. Set the scale with 0 at one end of the line being measured
C. Read the mark at the opposite end of the line to obtain the line length
D. Inches can only be estimated between footmarks

12' line @ 10 scale

12' mark

Engineer's scale

Figure 2-3 *An architect's scale, with fractional divisions (above), and an engineer's scale, with divisions in multiples of 10 (below).*

Measurement using an engineer's scale:

1. Identify the scale of the drawing to be measured.

2. Locate the scale that matches the drawing being interpreted. The engineer's scale places the number of the scale at the end of the instrument (Figure 2-3, step D).

3. Identify a line to be measured.

continued

4. Place the zero mark from the correct scale at one end of the line being measured (Figure 2-3, step E).

5. Read the markings along the scale to determine line length. Each mark on an engineer's scale represents one foot (Figure 2-3, step F).

6. Inches can only be estimated on an engineer's scale. Observe where the line ends between the marks and make an estimate of inches (Figure 2-3, step G).

PREPARING A LANDSCAPE PLAN

The landscape plan you need for your project should show all existing and proposed improvements you want to include in your landscape project. The landscape plan will assist you in pricing, locating, installing, and obtaining permits. Putting your ideas on paper before you begin work on the actual project will enable you to identify problems and challenges before you have invested significant time and money in construction. If important issues cannot be resolved during the process of preparing the plan, seek the assistance of a design professional.

To create a landscape plan, you will need to prepare a scaled graphic plan (called a **basemap)** of the site where the project is proposed. The basemap will locate existing improvements in the project area, so that the landscape plan can be designed around the important elements that you plan to retain. The landscape plan for your project can then be designed directly on top of the basemap by using sheets of translucent sketch paper. Several different ideas can thus be drawn and compared while preserving the basemap. Translucent drawing paper called **vellum** can be purchased in pads and rolls at hobby stores, art and graphic supply stores, and many college bookstores.

To obtain measurements for preparing this scale basemap, use the **baseline** method. This method requires establishing a reference baseline near the project that will be used for measuring and locating various project elements. Baselines can be found on the site using existing improvements, such as fences, walks, and walls. Baselines can also be created by placing posts at either end of the project and establishing a line between the posts. Regardless of which method you use, it is important that the baseline be straight, that it be as long or longer than the entire project, that it will not be disturbed by construction, and that it can be located again later in the project. These conditions allow measurements to be made from the baseline at any phase of the project, whether during planning or during construction. Once a baseline is established, measurements can be made down the baseline and then at a right angle to the baseline to locate objects on the site.

Preparing a basemap of your site

Time: 2–4 hours (longer if there are many objects to be measured or the site has sloped terrain or several obstructions).

Level: Easy (11 steps).

Note: You should either complete the project for using architect's and engineer's scale to measure drawings or know how to use scales before beginning.

Tools Needed:

1. 25' measuring tape.

2. 50' measuring tape.

3. 50 surveyor's flags. It is best to use white flags, to avoid confusion with the colored flags used by utilities.

4. 2 garden hoses.

5. Clipboard or portable drafting table.

6. 8.5" x 11" plain paper (for large sites you may want to use larger paper, either 11" x 17" or 18" x 24").

7. Pencil.

8. Architect's and engineer's scales.

continued

No materials required.

Directions:

1. Using the surveyors flags, mark the approximate location of your project and other key objects on the site (trees, buildings, ends of walls, edges of patios, and so on); see Figure 2-4, step A. Mark the corners of square and rectangular improvements and the ends of the straight improvements. Use the garden hoses to lay out any curved lines that must be measured, and then flag points along the curve.

2. Identify a baseline that is straight and passes through (or beside) your project area. You may use a wall, walk, or fence as long as it will not be disturbed by the construction. Mark a beginning point at one end of the baseline that is outside your project and an ending point that is beyond your project (Figure 2-4, step B).

3. Lay the 50' measuring tape along the baseline with the 0 marking set at the beginning point (Figure 2-4, step C).

4. On your sheet of paper draw a line along one edge that represents the baseline; if your baseline is in the center of the project, center the baseline on your paper (Figure 2-4, step D).

5. Using trial and error with each of the scales, determine which scale will best fit the length of the baseline on the paper. If drawing in one scale extends the baseline off your sheet of paper, switch to a smaller scale until you find one that will allow the line to fit on your paper (Figure 2-4, step E).

6. Locate an object on your site by measuring down the baseline from the beginning point to a point along the baseline that is at a right angle to the object. Use the 25' measuring tape to find the distance of the object from the baseline (Figure 2-4, step F).

7. Plot the location of the measured object onto the sheet of paper by measuring down the baseline the distance measured in step 6, then at a right angle to the baseline the same distance measured from the baseline in step 6. Place a

A. Locate improvements (white flags)
B. Locate baseline (gray flags)
C. Lay out tape for baseline

G. Plot improvements on plan

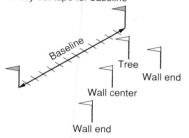

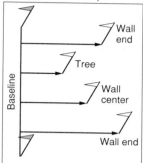

D. Draw baseline on paper
E. Determine scale by trial and error

H. Draw improvements

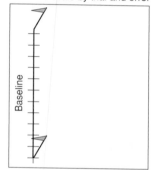

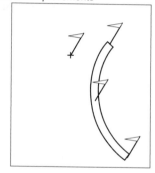

I. Add symbols for improvements
J. Label

F. Locate improvements

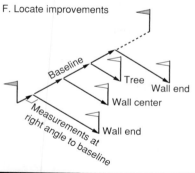

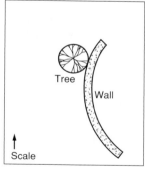

Figure 2-4 *Steps in preparing a basemap.*

continued

mark on the paper indicating the location of the object. Be certain all measurements on the paper are drawn using the determined scale.. If an object falls beyond the edge of the paper, redraw the baseline and objects using a smaller scale (Figure 2-4, step G).

8. Working one flag at a time, repeat steps 6 and 7 for all objects you flagged in step 1.

9. After finding the location of all flagged objects on the sheet of paper, connect the points with lines to show objects such as fences, patios, walls, and so on. (Figure 2-4, step H).

10. Symbols can now be added to show trees, shrubs, patios, and other objects (Figure 2-4, step I).

11. Label the objects and add a drawing title and the scale (Figure 2-4, step J).

ADDITIONAL METHODS FOR LOCATING OBJECTS

If baseline measurement is not appropriate to your project, alternatives exist. The methods described below are techniques that could be useful if you only need to locate a single item, or if using baseline measurement does not adapt well to your site.

To locate a single item on a site (e.g., a tree, a light pole, or the center of a pond) one of several shortcuts can be used. One method is to simply measure out from the side of a structure, particularly if there is a landmark on the structure, such as a door or window, to guide the measurement. A second method for locating a single point is to **triangulate** a location. If two known points exist (e.g., the corners of a building), measurements can be made from both points to locate an object. From each of the known points (A and B in Figure 2-5), measure the distance to the point to be located. On the base map, use the scales to duplicate the measurements. Draw arcs for each measurement from known points A and B. The point at which the arcs intersect is the unknown point (Figure 2-5). This process can be reversed to locate an object in the field, beginning from the scale drawing.

To locate an object at a right angle to a line, the **3,4,5 triangle method** will provide a greater level of accuracy than estimating. The steps for performing 3,4,5 triangle layout (math majors will recognize this as an application of the Pythagorean theorem used in triangle formulas) are shown in Figure 2-6. Along the line from which you are turning the angle, mark the point from the point at which a right angle is to begin (this may be the corner of the project or a line to locate an object). From that point measure 3' down the line from which you are turning the angle and make a second mark. From the beginning mark, anchor (or hold) a tape and stretch it in the direction you want the right angle to head. Anchor (or hold) another tape at the second mark you made. Adjust both tapes until the 4' mark on the first tape and the 5' mark on the second tape intersect. Mark directly below that point. A line traveling from the beginning point marked and passing through the point where the tapes intersect will be at a right angle to the first

A. Measure the distance to unknown point from known Points A and B.
B. Draw arcs for both distance A and B. Unknown point is where arcs intersect.

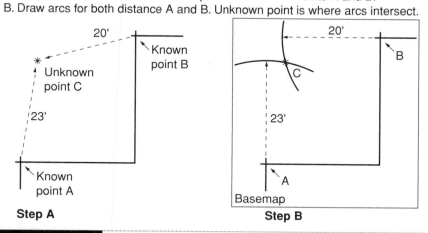

Step A

Step B

Figure 2-5   *Locating objects using triangulation.*

edge. When squaring decks or paving, you may mark on the forms or joists, and this technique will only require using the tape to make the 5' measurement. To improve accuracy, multiples of 3, 4, and 5 may be used (e.g.: 6, 8, and 10).

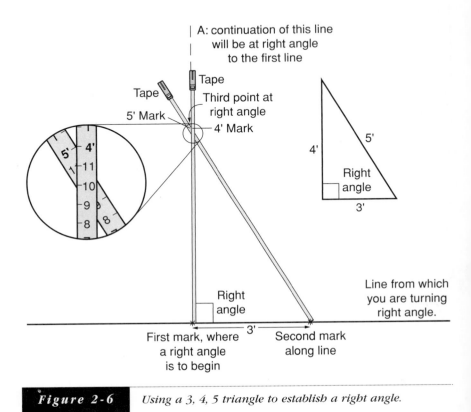

Figure 2-6   *Using a 3, 4, 5 triangle to establish a right angle.*

If the idea of preparing a landscape plan using blank paper seems impossible to accomplish, consider drawing on gridded paper. Many suppliers that sell scales and vellum also sell paper that has a background grid. This grid, often in increments of 4, 8, and 10 squares per linear inch, can be used to substitute for use of a scale in measuring. This may prove handy when attempting to jot down locations of objects measured in the field. Rather than using the architect's or engineer's scales, use the grids to indicate scale. For example, using 4 squares per linear inch would indicate a one-quarter–inch-per-foot scale, and so forth with other gridded paper. Base items and improvements can be drawn using either the grid or the matching scale (Figure 2-7).

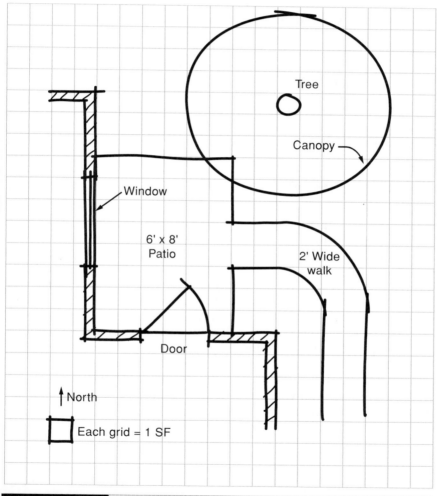

Tree

Canopy

Window

6' x 8'
Patio

2' Wide
walk

Door

↑North

Each grid = 1 SF

Figure 2-7   *Using gridded paper to locate improvements on a site.*

FINDING UTILITIES ON YOUR SITE

Construction activities that damage utility lines serving a site introduce several potential risks. Personal injury or death can occur from electrocution, explosion, or fire if utility lines are disrupted. Also, repairing damaged utilities may be quite costly, and the person who caused the damage may be required to pay for such repairs. In addition, your project may suffer now or in the future from repair activities. Take particular care when working with and around electrical, gas, and communication utilities; these pose the greatest risk to the homeowner if damaged during construction. Methods are available to help you in locating public utilities (gas, phone, electric, and others) that serve most structures and properties, but there is also the potential that private utilities (septic tanks, wells, exterior outlets, and so on) exist on a site that cannot be located. Regardless of whether these lines or structures are marked or unmarked, always dig with caution to reduce the chance of striking and damaging them accidentally.

Public Utilities

The risk of damaging utilities can be substantially reduced by calling for utility locations prior to beginning your work. Any time your plans require disturbing a site, contact utility companies before beginning construction, so that they can advise you of the location of their lines. Most utilities require 48 to 96 hours advance notice for locates and do not charge for locating services. To coordinate the task of locating utilities, many areas of the country have instituted One-Call service. One-Call is a toll-free phone number (1-888-258-0808) that provides the caller with the local number to contact for locating at least a portion of the utilities located on a site. The local number you call will *not* contact all utility companies that may have lines on a site but it should be able to contact at least public utilities servicing a property. You will still be responsible for verifying which utilities are not contacted.

When laying out a project, the locations of key improvements can be completed before calling for a utility locate. As long as markings are complete and accurate, this will help the utility company to safely identify their lines. To aid in identification, use white flags or paint to locate proposed landscape improvements, in order to avoid confusion with utility markings. Standard colors have been adopted by utility companies for identifying various types of utilities; the color-coding system is as follows:

- Red: electrical
- Yellow: gas
- Orange: phone, communication
- Blue: water
- Green: sewer

After utility lines have been located, carefully hand dig on either side of the utility marking. It is possible that the utility is slightly mis-marked, and hand digging will reduce the chance of disrupting the line. Some utility companies place a locator tape or cable with or slightly above the actual utility line to aid in locating it and to warn that it is close. If a utility cannot be located directly under its mark, dig a shallow trench perpendicular to the marking. Carefully and gradually dig the trench deeper to aid in locating lines that are not under the marking. Work the soil gently and in small amounts. Avoid chopping into compacted soil or using your shovel as a lever to pry up large sections of soil.

Locating utilities on your site

CAUTION

Locate all utilities on a site prior to starting any project. Failure to call for utility locations before construction begins could lead to injury, death, or financial loss.

Time: 1 hour.

Level: Easy (6 steps).

continued

Tools Needed:

1. Telephone

2. Writing materials

3. Landscape plan for your project

No materials required.

Directions:

1. Obtain the local One-Call number for the area in which work is planned by calling toll-free 1-888-258-0808.

2. Before contacting your local One-Call, gather the following information; the operator at One-Call will ask you for it:

   a. Your address and the specific location of the project.

   b. The type and extent of the work.

   c. Your name and address.

   d. The starting dates and times for the construction.

   e. A phone number One-Call can contact with responses and questions.

3. Place your call at least 96 hours (4 days) before beginning construction.

4. Contact your local One-Call with the above information.

5. Obtain a job number from the One-Call operator.

6. Survey the site to determine whether additional utilities may be present that would not be located by One-Call services. You may need to contact other local utility installers in order to identify and locate these utilities.

In certain situations the utility company may choose to have a representative present while work around their lines is being done.

Private Utilities

The site should also be reviewed for utilities that may not have been marked. Some sites include power, water, telephone, and other utilities installed by owners with no location recorded. Lines privately installed do not show up on utility company records and thus *are not marked by locators*. Look for outbuildings with power, yard lights, wells, gas tanks, septic tanks, or other site facilities that would obviously have some sort of connection with a utility line. Inspect inside buildings, particularly in the basement, for utility lines leaving a structure. Check the buildings on site for electrical breaker panels. Shutting off service at any unknown or unmarked breaker may stop service to areas served by those controls but may also reduce the chance of accidental shock.

You can also rent utility-locating equipment to help find private lines. Instructions for locating equipment are critical, because connections are often made to the utility line to trace its location. Lacking any other information to indicate the location of a utility line, you may have no choice but to unearth the entire line from its source to its end in order to find the location.

LAWS THAT MAY AFFECT YOUR PROJECT

Control over construction projects is typically expressed through the use of **ordinances** that control zoning and building. These regulations place stipulations on the use of land and the construction of improvements on property that lies within their legal jurisdiction. In large communities these regulations can be extensive and complex, and in small communities they may not exist at all. In the absence of an ordinance and permit system, a governmental unit may require that you adhere to one or more building codes. Some of the local government controls that are of concern are listed below. Contact local building officials, zoning administrators, the city forester, or the homeowner-association officer to determine whether any regulations apply to the work you are planning.

Legal Issues

Although you will not be performing work on your property as a contractor or design professional, you should be aware of some specific legal tenets that even the homeowner must follow. Failure to understand these tenets may result in your finding yourself spending more time in the courtroom than on your new patio! The first of these legal issues is trespass. Trespass is entering another's property or damaging that property. Be certain to ask permission if your work requires you to go beyond your own property, and avoid changing the direction of water that drains off your site. A second issue is nuisance. If you like to start work early, work late, or create a lot of dust or noise, you may be creating a nuisance to your neighborhood. Avoid unnecessary disruptions of the peace and quiet in your residential area to protect yourself on this count.

If your work creates situations that may harm or injure a person, use good judgment to avoid creating hazards. Negligence is defined as failure to take reasonable and prudent action to protect those around you and those who might pass or enter your site. Leaving holes unprotected in the evening, pruning over a public walk without watching for traffic, and similar actions that put others at risk should be avoided. Lastly, avoid construction activities that may cause neighboring properties to collapse. This concept, termed lateral rights, states that neighbors are entitled to the **elevations** that currently exist at their property line. For example, you cannot use a neighbor's property to taper the grade down to your new driveway without previously getting his or her permission.

The best policy for avoiding such problems is to do your research thoroughly and to communicate with the parties that will be affected by your work. There are many laws related to construction and landscaping, and this section serves as only an introduction to some areas that may be of concern. Talk with neighbors and government officials before beginning your project to determine whether they have any input. It is far better to hear their issues before beginning your project rather than later in legal documents. As with all issues that could be dangerous or damaging, contact your family attorney if in doubt.

Zoning Ordinances

Zoning ordinances control how land may be used. This type of control may restrict the type and dimensions of a structure one may place on a property and the uses of that structure. Also common in a zoning ordinance are requirements for setbacks, or distances between improvements and property lines. Parts of a zoning ordinance may control fences (such as a requirement to install them around a swimming pool, the height they may be, and materials that may be used to build them), deck locations and materials, water-feature or walkway construction, and railing requirements. You will need to determine whether there is a zoning ordinance in effect for the physical location in which you are working, and, if so, review the ordinance to find out how it may affect your project.

Tree Ordinances

Many cities have regulations on the types of plants that may be planted or removed. A street-tree ordinance will clarify the species that are acceptable for urban planting and may offer guidelines for plant location, construction of planters, maximum ultimate size of plants, and minimum size of plants to be installed. In addition to city ordinances, utility companies may also have requirements on plant- ings if there are overhead wires present where you plan to work.

Deed Restrictions

Planned communities, condominium housing complexes, historic dis- tricts, and occasionally entire communities may have restrictions that control what types of exterior improvements can be planned and built. These restrictions are intended to create uniformity of develop- ment or to protect historical authenticity. Review plans with authori- ties in these situations to verify the guidelines that construction work must follow.

Building Permits and Landscape-Plan Review

In order to verify compliance with ordinances, communities and coun- ties often require that a building permit be obtained prior to beginning construction. The process for obtaining such a permit may range from having an official review your ideas to attending a series of committee meetings in a process called landscape-plan review. Either method will most likely require preparation of a plan that contains descriptions of the work, along with elevations, or project grades, and dimensions. City building and zoning officials typically supervise the issuance of building permits and submissions for landscape-plan review. These same officials may also make inspection visits to the site to verify that the work is proceeding according to the submitted plan.

Checking regulations

Time: 1 hour to several weeks. This time is based on when the appropriate boards decide to take action on your project.

Level: Easy (4 steps).

Tools Needed:

1. Plan of your project.

2. Phone book with city, county, and other local government department phone numbers.

3. Telephone.

4. Pencil.

5. Paper.

6. Any permits, applications, surveys, or other documentation for your project that may be needed.

No materials needed.

Directions:

1. Identify the jurisdiction in which your project is located (condominium association, historic district, city, county, etc.)

2. Review the phone book for governmental offices that might have control over building projects. Look initially for a zoning administrator and then for a planning department. You may also be in a condominium homeowners association that controls exterior improvements or a historic district that will determine the form that projects can take. Identify the officers or committees that make these decisions.

3. Call the numbers listed for the above officials. Explain what your project is and ask to speak to the individual who administers the regulations for your type of work. If the respondent is not the correct official, he or she may know who is responsible for landscaping work and be able to direct you.

4. Discuss your project with an official who does have responsibility for determining whether you are required to obtain any permits, legal permissions, or easements, and whether your plan is subject to any reviews and inspections. If so, complete and submit all paperwork required for approval of your project.

It is important that you do not begin work until all approvals are in place. Be prepared to wait some time for the completion of the various steps of the approval process, and possibly to pay some fees, particularly if your project involves major work or activities that require permits.

PERFORMING CALCULATIONS FOR LANDSCAPE PROJECTS

Regardless of the project you are planning, math skills will be necessary to complete the project. Whether you are ordering materials, establishing grades, laying out a walkway, or performing any one of the many activities required of a project, you cannot avoid the need to perform math operations. Before starting these operations, it is necessary that you be able to perform basic calculations, calculate averages, and read formulas. If you are competent in these skills, the calculations for construction are typically a matter of making measurements and using those measurements to determine quantities. The primary math operations used in landscape construction include item counts, measuring lengths and perimeters, calculating the area of basic shapes, calculating volumes, and converting measurements to weight.

Landscape items such as benches, lights, plant material, and other **amenities** are measured and ordered by the quantity of each item used. An **item count** requires only that you perform a "head" count of each separate item being used. **Linear measurements** are used to calculate quantities for items that are purchased by length, such as edging and fence. Expressed as linear feet (LF), the linear measurement is also the building block for area and volume calculations. Linear measurements of edging, and materials placed around a standard geometric shape, can be obtained by calculating the **perimeter** of an object.

Materials such as brick, seed, and sod are generally purchased based on the area of the space they will cover. Quantities of materials necessary to cover flat surfaces require **area calculations**. Expressed as square feet (SF) or square yards (SY), the measurement of area requires recognizing the shape (or collection of shapes) of the space to be covered, and then calculating the area of that shape (or shapes) from linear measurements of key dimensions such as length, width, and radius. To calculate the area of an irregular shape, break the shape down into a collection of standard geometric shapes that approximately cover the entire irregular shape. Calculate the areas of each of these standard shapes and add them together to obtain the area of the total irregular shape.

For materials that are purchased in bulk, a **volume measurement** is typically required. Volume measurements are expressed as cubic feet (CF) or cubic yards (CY) and require an area measurement and knowledge of the depth of the material in order to perform calculations. Certain bulk materials may also require a conversion to a weight.

Math and project measurements can be can be simplified if you use the decimal system rather than inches and fractions. Most math functions benefit from converting numbers to feet and tenths of feet.

Performing Item Counts

Item counts are the simplest of the calculations to perform. Locate and count the number of symbols representing the object for which you need a quantity.

Performing Linear Measurements

Identify the two points between which a measurement is required and measure between the points to obtain the dimension. If you are using a plan, use the proper scale to measure between the points. Repeat the process for all items that require linear measurement.

Performing Perimeter Calculations

Identify the shape of the perimeter you need to calculate in Figure 2-8. If you are working with an irregular shape, the fastest method for measuring a perimeter may be to take a direct measurement. Make any linear measurements necessary to complete the formula, and then calculate the perimeter using the appropriate formula.

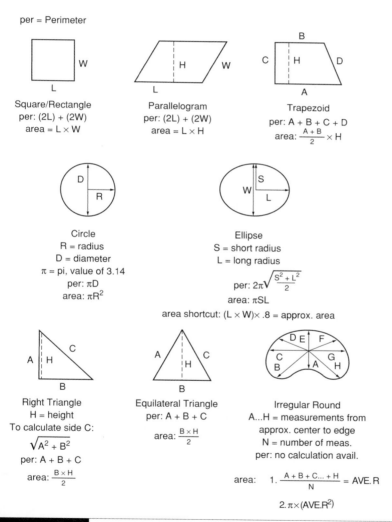

per = Perimeter

Square/Rectangle
per: (2L) + (2W)
area = L × W

Parallelogram
per: (2L) + (2W)
area = L × H

Trapezoid
per: A + B + C + D
area: $\frac{A + B}{2}$ × H

Circle
R = radius
D = diameter
π = pi, value of 3.14
per: πD
area: πR²

Ellipse
S = short radius
L = long radius
per: $2\pi\sqrt{\frac{S^2 + L^2}{2}}$
area: πSL
area shortcut: (L × W)× .8 = approx. area

Right Triangle
H = height
To calculate side C:
$\sqrt{A^2 + B^2}$
per: A + B + C
area: $\frac{B \times H}{2}$

Equilateral Triangle
per: A + B + C
area: $\frac{B \times H}{2}$

Irregular Round
A...H = measurements from approx. center to edge
N = number of meas.
per: no calculation avail.
area: 1. $\frac{A + B + C... + H}{N}$ = AVE. R
2. π×(AVE.R²)

Figure 2-8 Perimeter and area formulas for common shapes.

- Square/rectangle/parallelogram: Shapes with two sets of parallel sides. Squares and rectangles have right-angled corners, and parallelograms have no right-angles corners.

- Trapezoid: A shape with one set of parallel sides and two right-angled corners adjacent to each other at each parallel side.

- Circle: A round shape with edges equidistant from the center.

- Ellipse: A rounded egg shape with edges variable distances from the center.

- Triangle (right or equilateral): A shape with three sides. A right triangle has one right angle, and an equilateral triangle has all three sides of equal lengths.

- Irregular Round: Irregularly or kidney-shaped circles for which a center point can be located.

Performing Area Calculations

Evaluate the shape for which you need to calculate the area. Determine whether the shape is an easily measurable single shape (as described in perimeter calculations) or a collection of geometric shapes. For standard geometric shapes, make the measurements required for each area formula (Figure 2-8). If your shape is irregular, you will most likely be able to break that shape down into standard shapes. Draw these shapes inside the irregular shape, covering as closely as possible the majority of the area being measured. After the irregular shape has been divided into regular shapes, measure each shape and calculate the area, and then total the answers to obtain the area for the shape. Be certain all calculations are correct, and if only a portion of the shape is utilized (such as one-quarter of a circle), reduce the total appropriately.

If necessary, convert your answer to the appropriate unit. Some conversions applicable to area measurements:

- If the measurement required must be expressed as square yards, divide the square-foot total by 9 to obtain square yards.

- To obtain the number of acres, divide the square footage by 43,560.

- To obtain the number of squares of sod required (each square being 100 SF), divide the square footage by 100.

- To obtain the number of rolls of sod (each sod roll being 1 SY), divide the square footage by 9.

Volume Calculations

Begin volume calculations by determining the square footage of the area that is to be filled. Determine the depth of the layer to be filled and place the numbers in the following formula:

$$\frac{\text{area (in SF)} \times \text{depth (in inches)}}{12} = \text{CF}$$

(Note that the depth is not converted to feet but left in inches.)

Cylinder Volumes

If the volume to be calculated is a cylinder, use the following formula:

cylinder area × cylinder height = CF
(both area and height must be expressed in the same unit,
preferably as feet or decimal portions of feet)

Berm Volumes

If the volume to be calculated is for a landscape berm with even side slopes, use the following formula:

(berm length × berm width × berm height) × .7 = CF
(all berm measurements must be in feet or decimal portions of feet)

If necessary, convert your answer to the appropriate unit. Some conversions applicable to area measurements are:

■ If the measurement required is in cubic inches and must be converted to cubic feet, divide the cubic-inches total by 1,728 to obtain cubic feet. (12" × 12" × 12" = 1,728 cubic inches)

■ If the measurement required is in cubic feet and must be expressed as cubic yards, divide the cubic foot total by 27 to obtain cubic yards. (3' × 3' × 3' = 27 cubic yards)

Weight Conversions For Bulk Materials

To convert volume measurements to bulk measurements, a conversion factor is required. Multiply the CY or CF from your volume measurement by the conversion factor for the material you are using:

■ 1 CF soil, dry and loose = 90 lbs.

■ 1 CF soil, moist = 75 to 100 lbs.

- 1 CF sand, dry = 100 lbs.

- 1 CF limestone, uncrushed = 160 lbs.

- 1 CF concrete = 140 lbs.

- 1 CY soil = 1.2 Tons (2400 lbs.)

- 1 CY concrete sand = 1.5 Tons (3000 lbs.)

- 1 CY angular crushed stone (class 5 aggregate, 1" roadstone, or equivalent) = 1.25 Tons (2500 lbs.)

To illustrate the math concepts just explained, the example in Figure 2-9 shows how the various measurements (from formulas shown in Figure 2-8) are used. Building on each previous measurement, measure the length of the patio and then multiply it by the width to calculate the area. Then use the area to calculate the volume for the patio base material.

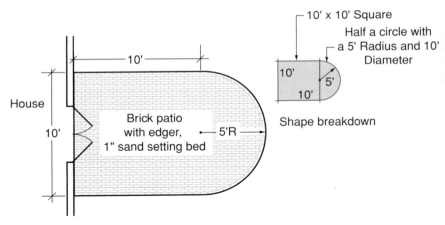

Edger perimeter: Length of two straight 10' sides plus curved edge
(Note: no edger against house)
(10' + 10') + [(3.14 × 10)/2] = Perimeter
20 + 15.7 = 35.7 LF of Edger

Brick area: 10' Square plus half of 5' radius circle
(10' x 10') + [(3.14 × 5²)/2] = Area
100 + 39.25 = 139.25 SF of brick

Sand volume: 1" of sand under 139.25 SF of brick.
(139.25 × 1)/12 = 11.6 CF of sand.

Figure 2-9 *Example of math calculations. The steps illustrate the calculation of perimeter, area, and volume for an example patio.*

Math for construction

CAUTION

- These methods of calculation have the potential for error, so if material orders are critical, a more precise method of measurement should be selected.

- Seek supplier assistance for calculating amounts of concrete, stone, and other expensive materials.

Time: 1–2 hours.

Level: Moderate (2 steps). Calculation required.

Tools Needed:

1. Calculator with basic functions, square, and memory.

2. Paper.

3. Pencil.

4. Architect's and engineer's scales.

5. Landscape plan on which to perform calculations. You can use any plans you have prepared, or plans found in magazines that are drawn to scale. A designer may also be willing to give you an old plan.

No materials needed.

Directions:

1. Search through the landscape plan and list all landscape elements that require item counts, linear measurements, area measurements, and volume measurements.

2. Using the directions for each type of measurement, perform the calculations necessary to determine the items, lengths, areas, and volumes as shown on the plan.

continued

- Perform an item count on all items that are purchased individually, such as light fixtures, benches, pools, and similar types of items.

- Locate any edge restraint or fencing that requires linear measurements.

- Find bedding areas, paved areas, turf areas, and other spaces that will require area measurements. Determine whether they match standard geometric shapes. Perform the linear measurements necessary, and then calculate the area.

- For areas that require volume, calculate the area and use the thickness of the installation to determine the volume.

- If required, convert any volumes to weight.

PRICING YOUR PROJECT

Of all the factors that might halt a project, few will stop progress as quickly as not having the funds necessary to complete the work. To determine project expenses, materials costs can be calculated using the math procedures from the previous section applied to your landscape plan. Performing a **materials take-off**, or a calculation of the quantity of all materials to be used for the project, is the first step in pricing your project. Multiply the quantities of the take-off by the unit costs obtained from suppliers to determine the total estimated cost of the project.

When performing a material take-off, consider slightly increasing the actual quantity required for certain materials. Landscaping projects typically use many materials that may settle during shipping, be blown away, eroded, or otherwise lose "bulk" from the time they are first loaded to the time they are placed on the site. To compensate for this loss, it is not uncommon to add 10%–15% to bulk-material orders to ensure you have adequate materials to complete the job. Materials that would typically be treated in this manner include soil, sand, base material, and other bulk granular materials. Bulk materials are not the only materials that should be over-ordered to compensate for construction conditions. Unit pavers and wall materials typically experience waste as high as 5% when splitting is required by the design. In complex projects the waste percentage can be much higher. However, materials calculated using an item count should not be over-ordered.

This concept of over-ordering may require that excess material be removed from the site, but this expense is substantially less than the cost of additional deliveries. This approach also avoids the annoyance of being "just short" of having enough material to finish. Maintaining a small stockpile of materials that can be used for repair and maintenance will be helpful later when attempting to match paver colors or wall materials.

Estimating the cost of your project

Time: 2–4 hours, depending on the complexity and scope of the design.

Note: This project is a calculation of material costs only. No labor or contracting charges are included in this calculation. Estimates may be required for materials in projects described later in this book, so if you are unsure of any terms please check the glossary or complete this exercise after you have planned your project.

Level: Moderate (5 steps). Calculations required. Complex task.

Tools Needed:

1. Calculator.

2. Paper.

3. Pencil.

continued

4. Engineer's and architect's scales.

5. Landscape plan of project. Use one you have prepared or a scale drawing from a magazine or designer.

No materials required.

Directions:

Design Review:

1. Review the design and identify and list all landscape construction work segments required by your landscape plan. To keep this task organized, group tasks according to the headings under the material take-off step below. For each task, list all materials, even minor products necessary to perform that portion of the work. Your project may not have all the aspects shown below.

Materials take-off:

1. Identify the scale of the drawing, make your measurements, and then calculate the quantities of each material identified in the previous step. Measurements will be linear, perimeter, area, volume, and weight. Calculate the quantities of materials required for each portion of the project.

   ■ Site preparation and/or demolition. Locate any area that will be disturbed by construction, particularly those that require the removal of existing materials. Measure and calculate:

   • Removal of trees and shrubs (typically calculated for each item removed).

   • Removal of sod and scrub plant material (typically calculated by the square foot).

   • Removal of undesired paving, typically calculated by the SF.

- Grading and erosion protection. Identify any location at which the site is disturbed and/or the finish grade differs from the existing grade. Measure and calculate:
  - Purchase and distribution of additional soil (typically calculated by the cubic yard (CY).
  - Installation of **erosion mat**, calculated by the SF.
  - Installation of **drainage tile** (calculated by the linear foot (LF).
- Wall construction. Locate all walls, identify the materials used for construction, and determine the dimensions; then:
  - Measure each wall installation, separated by different materials. Calculate the square footage (SF) of the face of the wall, including the portion buried below grade.
  - Measure the length of all walls and calculate the amount of capstone that may be required (typically calculated by LF).
  - Calculate the quantities of any materials placed for base below the wall or as fill behind the wall (typically calculated by the ton or CY).
  - Calculate any drainage materials needed behind the wall (typically calculated by LF).
- Paving installation. Locate all paved areas, identify the materials used for construction, and determine the dimensions; then:
  - Measure each paved area separated according to paving type (typically calculated in SF).
  - Determine the base and setting-bed materials that are necessary (typically calculated by the ton or CY).
  - Calculate any edge restraint necessary (typically calculated by CF).

continued

- Wood construction. Identify all wooden structures on the site; then:
  - Identify the wood members used in the design.
  - Perform a count of the number, size, and length of each member required.
  - Calculate the connectors and fasteners needed for the project. Each individual piece should be counted.
- Fencing or free-standing wall construction. Identify all locations at which fences or free-standing walls are located, and the dimensions and materials for each installation; then:
  - Calculate the face area of free-standing walls (typically by SF)
  - Determine the fencing quantity (typically by counting each piece of material required).
- Amenity installation. Locate each site **amenity** required by the design:
  - Perform an item count for each different amenity. Group according to similar amenity types and styles.
  - Include quantities of materials required to install, anchor, or support the amenities.

2. Visit a local lumber yard, garden center, or hardware store to get the prices for each item. Ask the suppliers and dealers for their advice on material choices and quantities.

3. Multiply the price by the number of each item used in the project.

4. Add any additional charges you may incur, such as delivery, taxes, special tools and rental equipment you will need to complete the project.

5. Add the unit costs and additional charges to get an estimate of your project cost.

PROJECT LAYOUT

When all conditions indicate that the project can proceed, the beginning of physical work occurs when you locate your project on the site. This step provides the information necessary to identify exactly how your wall, pool, fence, or other project will fit into the existing landscape. Laying out your project will also allow you to make **field adjustments**—minor changes before construction that might save a tree, avoid a utility, or improve the design.

The process of project layout can take a number of forms, but the simplest for the homeowner is to use the same baseline used in preparing the basemap. When the landscape plan is complete, the baseline used to locate existing objects can be reestablished to lay out the project. If the basemap was accurately drawn, the landscape plan that was prepared can be transferred to the site with an equally high degree of accuracy (Figure 2-10).

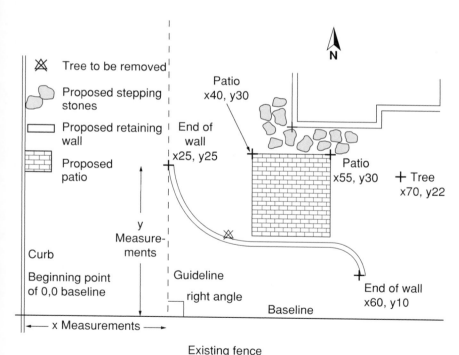

Figure 2-10 *Baseline layout of a plan. All points are located using x and y coordinates laid out in reference to the beginning point. The pairs of numbers indicate x and y measurements.*

Using a baseline to lay out your project

Time: 2-4 hours, depending on the complexity and scope of the project.

Level: Moderate (14 steps).

Tools Needed:

1. 25' measuring tape.

2. 50' measuring tape.

3. 50 surveyor's flags. White is the preferred color, to avoid confusion with the flags used to mark utility lines. You can recycle the flags from the basemap preparation.

4. Marking spray paint.

5. Architect's and engineer's scales.

6. Paper.

7. Pencil.

8. Plan for your project prepared using a baseline.

No materials required.

Directions:

To create baseline layout measurements for key points of a project:

1. On your plan, draw the location of the baseline used to measure and plan the project (this may be the same baseline as the one used to gather measurements for the basemap).

2. Identify the beginning point of the baseline.

3. On your plan, identify all key points of your project that will need to be located in the field. These points may include the corners of a patio, the beginning and end of a wall or fence,

or other key locations. Identify as many points as possible to get an accurate field location of the project.

4. Select one of the key points on the design to be measured.

5. Draw a guideline at a right angle to the baseline that passes through that key point.

6. Using the scale, measure along the baseline from the beginning point to the guideline. This measurement will be called the x measurement.

7. Measure along the guideline from the baseline to the key point. This measurement will be called the y measurement.

8. You should now have both x and y measurements for the key point you selected.

To locate that key point (from the previous steps) on a project site:

1. Locate the baseline on the project site. Use marking paint to locate the beginning and ending points. Place the 50' measuring tape along the baseline with the 0 marking at the beginning point. Use this baseline to find the x measurement.

2. Locate the x measurement for the key point along the baseline.

3. From the x measurement extend a line at a right angle to the baseline toward the key point being located.

4. Using the 25' measuring tape, locate the y measurement for the key point along the right-angled line. That measurement is the correct location for the key point you are trying to locate.

5. Place a flag at this location with a label for what the point represents (corner of a patio, end of a wall, etc.)

6. Repeat these steps of measuring and locating for all key points that require placement.

7. Paint between flags at corners of objects and points along a curve to find the location of edges between flags.

TOOLS FOR CONSTRUCTION

Completing the projects in this book will require an array of tools commonly used by homeowners and a few tools used by specialists. Most of the essentials are common carpentry and landscape tools, and specialized pieces of equipment can typically be rented. With a few exceptions, most reputable rental centers will carry items such as vibratory plate compactors and major power tools. Though not required for every activity, each of these tools will be used if all projects in this text are completed. See Figure 2-11 for illustrations of these tools. To best prepare for all situations, bring a basic toolset of a claw hammer, screwdrivers, pliers, tape measure, level, and utility knife to each construction project.

- Hammers (claw, 2 lb. sledge, large sledge).
- Screwdriver, either hand or power, with standard, Phillips-head, and square-drive bits.
- Electric drill.
- Auger and spade drill bits in 1/8" to 1.5" sizes, and standard, Phillips-head, and square–drive bits.
- Pliers.
- Saws and cutting tools (circular saw, carpenter's saw, hacksaw, utility knife, tin snips)
- Carpenter's and line levels.
- Ladder.
- Landscaping and excavation tools (wheelbarrow, rakes, shovels, post-hole excavator, drop spreader, hand tamper, fence-post driver).
- Masonry tools (magnesium float, rubber mallet, stone hammer, brick set).
- Pruning tools (pruning saw, loppers, bow saw).
- Marking and measuring tools (tape measure, chalk line, carpenter's square, marking paint).
- Large gardening equipment (vibratory plate compactor, sod roller, drop spreader).

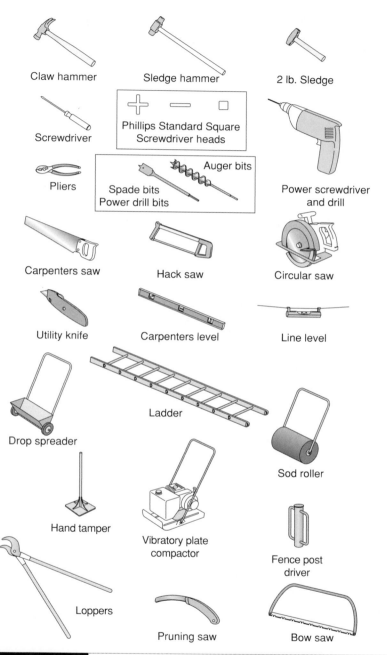

Claw hammer

Sledge hammer

2 lb. Sledge

Screwdriver

Phillips Standard Square
Screwdriver heads

Pliers

Auger bits
Spade bits
Power drill bits

Power screwdriver
and drill

Carpenters saw

Hack saw

Circular saw

Utility knife

Carpenters level

Line level

Drop spreader

Ladder

Sod roller

Hand tamper

Vibratory plate
compactor

Fence post
driver

Loppers

Pruning saw

Bow saw

Figure 2-11   *Tools used in landscape construction*

Chapter 3

PREPARING THE SITE FOR CONSTRUCTION

During my first job in landscape construction, I nearly panicked from the fear that once I had started preparing the site it would never look the same. After the first **skidsteer** passed over the lawn, ripping the turf to shreds, and the first tree to be removed was cut down, the feeling of "What have I done?" was overwhelming. As the project progressed and the fear subsided, the realization that to improve the site some impact would be necessary replaced the visions of spending the rest of my life piecing the site back together. Site preparation is the step designed to ease these initial project fears.

Few landscapes can weather the loss of a 50-year-old oak damaged during construction, and many sites will have plants and old projects that are in the way of your new plans. It is unusual for the site on which you want to make landscape improvements to be ready for immediate construction, so most projects will require that you complete some site-preparation before you begin building. Two important considerations for preparing a site include protecting it from damage during construction and removing unwanted elements. Although you may experience a temptation to try to cover or work around the intrusions, the price of preventing construction mishaps is often far less than the cost of repairing mistakes. Practicing proper preservation and removal of existing site elements throughout the construction process is a sensible choice.

PROTECTION OF PLANT MATERIAL

Of the many site elements that require consideration, the need to protect plant material should be obvious to a landscape enthusiast. Unfortunately, some people are not familiar with plant culture and fail to practice proper protection techniques when working around plant material. Construction poses several perils to the mature plant species at a site, and in most cases the damage does not become apparent until several years after the construction project has been completed.

Familiarizing yourself with the identification and cultural requirements of plants helps you recognize which plants are most sensitive to construction and which construction practices pose the greatest risk to plants. Damaging a tree or larger plant will require a long recovery period, if they are able to recover at all. Establish a goal of trying to keep each protected plant in its pre-construction environment throughout your project. This will help reduce overt and latent construction damage. If protection is not an option, some smaller plants may be dug up and stored in containers, then replanted when construction is complete. Other valuable plants may be permanently moved to new locations if they are not compatible with the new design.

Protection of plants must go beyond simply protecting the trunk and the foliage. The root zone of a plant is as important as the trunk and **canopy** are. The root zone of a plant often extends from the trunk to well beyond the **dripline** (Figure 3-1), or the imaginary line on the ground directly below the outermost foliage. In some situations the root zone may cover an area over twice that of the canopy. Compaction, trenching, mixing of chemicals, excavation, storage of materials, and related construction activities within the dripline of a plant can easily damage the root system of a plant, eventually leading to damage to the remainder of the plant. Protection must extend at least to the dripline of the plant, and in some cases you should consider restricting activities outside the dripline. Because of their size and vigor, many shrubs are more durable than trees when exposed to similar construction activities. Despite this, shrubs should be afforded the same protection given trees, to enhance their chances of survival. Because of the sensitivity of plants to damage, planting is typically one of the last steps in any construction project.

Not all activities are damaging to the root zone of a plant. Installation of bedding plants and ground covers can be undertaken with minimal damage to the roots of a larger plant. Some construction may have to

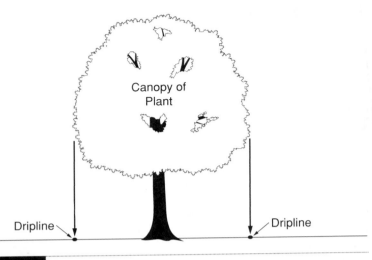

Canopy of
Plant

Dripline

Dripline

Figure 3-1 *Location of plant dripline.*

take place under the canopy of a plant, but the intent is to identify those areas that are not part of the construction and to protect them while minimizing the damage where work has to take place. In those areas in which equipment must operate under trees, consider heavy mulching or plywood "paths" to limit compaction.

Protecting plants from construction

Time: 1–4 hours, depending on the number of plants that need to be protected.

Level: Easy (5 steps).

Tools Needed:

1. Plan of your project.

2. Fence-post driver.

Materials Needed:

1. Utility fencing (orange woven plastic fence), 1 linear foot for every foot of perimeter to be fenced.

2. Fence posts, 1 post for every 10 linear feet of fence.

3. Plastic twist ties, 2 per post.

4. Lightweight rope, 10 feet per shrub to be bound.

Directions:

To protect plants from damage due to construction, complete the following steps:

1. On your construction plan, sketch in the dripline of any tree and shrub to be protected. If you are unsure about the extent of root growth, move your markings further out from the plant.

2. Review your construction plan for any areas that could be disturbed by grading, utility trenching, construction traffic, or other construction activities.

3. Identify where construction might unnecessarily be directed under the tree and shrub canopies. These areas should be protected.

4. Install fence posts every 10 feet around the perimeter of the area to be protected. Stretch the utility fence between the posts (Figure 3-2). Secure the fencing to the posts using the twist ties.

5. Shrubs near the construction area can be protected by bundling the **canes** (stems) together using a lightweight rope. Tie a loose loop with the rope around the base of the shrub. Slide the loop upward until the shrub canes are upright and secured.

continued

Figure 3-2
Fencing placed at dripline of a tree to protect the root zone.

REMOVAL OF UNWANTED SITE ELEMENTS

Within each project there are likely to be elements that do not fit the scheme of the plan. Whether these elements are **hardscape** (paving, fences, and so on) or **softscape** (trees, shrubs, sod, etc.), it may be necessary to remove them. For some the idea of taking out existing plants or hardscape creates undue tension, and they react by attempting to save every element in a landscape. Use good judgment in identifying which parts of an existing landscape are invaluable and still serviceable, but also recognize that on any site there will be some elements that have served their useful purpose. If your plan development is sound, you can identify and replace those items. Keep in mind that your ultimate goal is to produce a better landscape.

Softscape Removal

Techniques for removal of plant material differ, depending on the type and size of plant being removed. When removing plants it is important that the majority of the vegetative parts be removed. Uneven settlement will occur in paving or finish grades if large portions of stumps or excess amounts of plant roots are left in place to decay. Excavating all roots is impractical, but you should remove major roots near the stump, and remember not to bury plant-material waste in areas in which the finish grade is critical.

Preparing a site for planting or construction will first require removal of plants growing on the surface. Stripping sod and ground cover requires scraping a thin layer of plant growth off of a large area. This material can be cut and removed by hand if the area is small. Large projects, however, are usually stripped using sod cutters, skidsteers, or large earth-moving equipment that can quickly peel the ground cover away from the topsoil over extensive areas.

Large plants that are to be removed from a site will require techniques different than those for removing ground covers. You can remove larger specimens, such as trees and shrubs, individually and dispose of the debris or chip it into compost. Masses of large plants must be addressed one at a time. Before removing trees and shrubs, assess whether the appearance or spread of the plants can be altered by **pruning**. You may be able to save some trees and shrubs from removal by cutting a few key branches to raise the canopy or expose a view.

CAUTION

- Seek professional help for removal of any plant that is over 20' tall, that could cause injury or damage if improperly removed, or that is near overhead utility lines.
- Use caution when cutting and lifting.
- Follow the manufacturer's instructions when using any power equipment.

Removing sod and ground covers

Time: 1–2 hours; varies based on the size of sod area to be removed.

Level: Moderate (7 steps). Extensive digging required.

Tools Needed:

1. Plan for project.

2. Marking paint or garden hose.

3. Wheelbarrow.

4. Square-nosed shovel.

5. Disposal vehicle, home **compost** area, or area to reuse sod.

No materials needed.

Directions:

1. Identify on the plan the area of sod to be removed.

2. Use the paint or the garden hose to mark the perimeter of the sodded area.

3. Use the square-nosed shovel to make a vertical cut around the entire perimeter.

4. Plan the removal in parallel strips that are the width of the shovel blade.

5. Start removal of the sod at a high side of the project.

6. With your foot, push the shovel just under the surface of the sod. You should skim off the top two inches of sod and soil. Skim a small section of sod, then cut it off and place in the wheelbarrow. Continue this operation for the entire area within the perimeter (Figure 3-3).

7. Compost or dispose of the sod properly. If you have bare areas that could use sodding, transplant the excavated sod to these areas. Sod should be replanted within 4 hours of excavation, and watered daily until rooted.

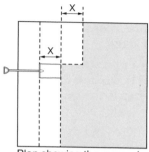

Plan showing the removal
in strips the same dimensions
as the shovel blade

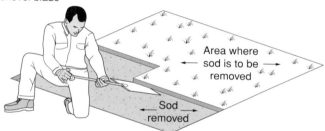

Figure 3-3 *Removal of sod.*

Hardscape removal

Like the plants on a site, there will undoubtedly be existing improvements (that is, manmade objects) that interfere with the design of your project. Some existing elements may be creatively reused in place, but not all will fit into your plans. There is a temptation to cover over existing site elements rather than remove them, but this practice is not recommended for the following reasons:

- If the covered element is a utility line, there will always be uncertainty as to whether it is still active.

- Buried pavements and foundations can restrict plant growth. Roots for trees and shrubs may not be able to reach groundwater levels, and sod may dry out faster over an old improvement left in place.

- Future excavation will be more difficult and expensive if it encounters old footings or slabs.

- Settlement of topsoil will occur at different rates over old site elements left in place than over soil that doesn't contain buried improvements, leaving uneven surfaces in lawns and paving.

- Removing pavement and other permanent improvements is a labor-intensive operation. Renting specialized demolition equipment for this phase of work is highly recommended if large areas of pavement need removal. Removing slab pavement usually requires that it be first broken into pieces and then hauled away. For small areas this operation can be done by hand with a sledge hammer or **pneumatic powered jackhammer**. Larger areas will require that the pavement be sawn into smaller pieces with a **cutoff saw** before breaking it apart. Work carefully in removing the old surfacing. Be careful not to force pieces of the pavement against structures, footings, or other improvements that need to remain intact. Obtain assistance in lifting and removing heavy pieces.

- Footings and foundations buried to frost depth will require the use of a **backhoe** to excavate a working area around the footing, then breaking and removing the footing using the backhoe bucket. A contractor typically has the equipment needed to complete these tasks.

Removing unwanted trees and shrubs

Time: 1–4 hours per plant.

Level: Challenging (16 steps). Specialized equipment operation and heavy lifting required. Assistance may be required.

Tools Needed:

1. Plan of project.

2. An old pair of **lopping shears** that you don't mind using to cut roots.

3. **Pruning saw**.

4. Lightweight rope, 10 feet for each shrub removed.

5. Vehicle for removal of debris

Directions:

Removal of small trees:

1. Identify on the plan any small tree that is not part of the project plan.

2. Using the lopping shears and pruning saw, remove as many small branches throughout the tree as you can safely reach from the ground. Open a path to the trunk of the tree.

3. Examine a circular area around the tree that is one and a half times the height of the tree to verify that it is free of objects that could be damaged by the tree falling on them. Verify that there are no overhead utility lines in the same circular area. If there are utility lines or objects within the circular area, seek professional help in **felling** the tree.

continued

4. Determine a safe direction for the tree to fall that also leaves you a safe route for walking away from the tree when it falls. This exit route should be behind and away from the planned direction of fall.

5. On the side of the trunk facing the direction you want the tree to fall, cut a V-shaped notch through one-third the diameter of the trunk.

6. On the side opposite the V notch, begin cutting from above the V notch in a downward direction toward the V (Figure 3-4). Stand to one side of the tree with your escape route behind you. Never stand directly in front of, or opposite to, the planned direction of fall.

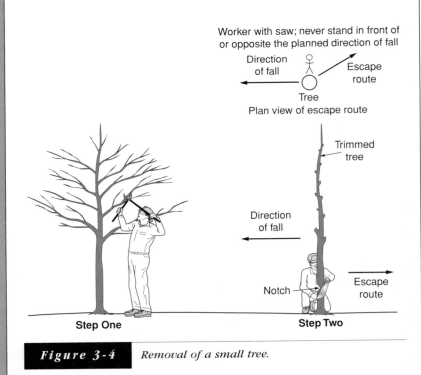

Figure 3-4 *Removal of a small tree.*

7. When the tree starts to fall, remove the saw and walk away along your exit route. Do not stand near the tree while it falls. When the tree hits the ground it may kick back several feet in a direction away from where it first struck the ground.

8. With the tree on the ground, finish removing branches and cut the trunk into manageable pieces; remove all debris from the root zone.

9. Dig a 2' diameter circle around the stump of the tree. Within that circle, use the lopping shears and pruning saw to cut and remove all roots encountered. Dig the remaining portion of the stump out of the ground.

10. Compost or dispose of waste properly.

Removal of shrubs:

1. Identify on the plan any shrub that is not part of the project plan.

2. Using the lopping shears, remove as many branches and stems of the shrub as possible. Shrubs with large numbers of small canes can be bundled up using a lightweight rope and cut at the base in a single operation.

3. If large branches or stems remain, remove them in 3-foot-long sections using the pruning saw.

4. Cut any remaining stem to the ground.

5. Dig a 2' diameter circle around the stump of the shrub. Within that circle, use the lopping shears and tree saw to cut and remove all roots encountered. Dig the remaining portion of the stump out of the ground.

6. Compost or dispose of waste properly.

RECYCLING AND WASTE DISPOSAL

Methods for landscape waste disposal include landfills, recycling, and creative reuse of by-products. Whether the method is environmentally sensitive or insensitive, landscape waste will be an issue for which all must prepare. Temptations to landfill all waste is now counterbalanced by the limitations of disposal space and restrictions on placement of organic waste in landfills. Most landscape waste materials can be recycled. Metals from fences, wood from decks, plastic and metal from edging, concrete and asphalt paving, crushed bricks, wiring and piping, and many other materials unearthed can be taken to various centers that will recycle the products. Many of the by-products of site preparation can also be used in other aspects of landscaping if you are willing to make the effort to reuse and recycle.

From the first bathtub shrine to your childhood tire swing, adaptive reuse of old materials in the landscape has been taking place for decades. Listed below are a few creative suggestions for reusing hardscape material:

- Stone wall material is timeless and can be reused as walls, paving, steppers, edging, or lining swales.
- Bricks or pavers can be reused as edgings or steppers.
- Concrete pieces can be used for wall material or crushed and spread as a loose aggregate surfacing. When doing this, ensure that all **reinforcing materials** (wire mesh, reinforcing bars, etc.) have been removed before reusing as a surfacing material.

Healthy, woody plant material can be recycled using a **chipper** to create shredded mulch. The quality of this mulch will vary, depending on the parent material, but most woody plant parts can be used for landscape beds, whereas material with small branches and foliage may be acceptable only for field mulch or composting purposes. Plant material that is diseased or insect infested should be disposed of rather than recycled. Sod that is stripped from the site can be composted, along with other plant waste from the site.

Clean soil without debris can be used as fill when properly placed and compacted. This filling procedure can be done on site or transported to an off-site facility. Soil contaminated with wood, paving, or other debris will deter trenching and future underground work in the fill area. The decomposition of large quantities of organic material, such as those created by stripping sod, will render the surface unstable.

The use of landfills for disposal of waste has limitations, including costs and disposal restrictions. Many localities have strict regulations on what materials can be placed in landfills, including limitations on plant material. Typical use of a landfill requires the owner to haul and place the waste material, paying fees based on the weight of the material disposed. Organic material, including green plant matter, is sometimes composted if the solid-waste facility has processing equipment.

Always secure waste and debris when traveling to a recycling or landfill facility. This will require using vehicles with tailgates or enclosed cargo areas, covers, or tying down loose plant waste. If you are unsure regarding regulations for proper hauling of waste, contact your municipal or state department of transportation.

Chapter 4

GRADING, EROSION CONTROL, AND DRAINAGE

Working with the topography on a site is one of the most interesting ways to alter a landscape. Building mounds, sculpting landforms, creating **drainageways**, and other creative ways to spice up a flat site work wonders in setting the stage for the **hardscape** (paving, decks, and so on) and softscape (plants) that follow. Of course, there are good reasons for doing this type of work, such as preventing floods and keeping the soil on the site, but if approached with the attitude that all aspects of grading work together to form a foundation for the landscape, your grading projects will take on new meaning.

Water and soil play leading roles in the successful landscape. Soil is important as a building base and growth medium. Water, in its various roles, can impact the landscape through drought, freezing, flooding, ponding, and plant growth. When combined on the surface of an exposed site, water and soil combine to create the potential for erosion. A homeowner's ability to properly manipulate these two elements will often determine the long-term success or failure of a project. Through proper grading, erosion control, and drainage, problems with water and soil can be limited.

SITE GRADING

Most landscape projects require moving soil to accomplish design goals. A simple concept of **grading** could be described as removing soil from where it is not needed and placing it where it is desired. Grading a site requires knowing the steps of the grading process, the ability to visualize landforms, and skill in estimating quantities of soil to be moved. For large projects these activities are engineered with a high degree of precision, but for smaller homeowner projects, grading using simpler means can be successful.

Grading is accomplished through a process called **cutting** and **filling**. Cutting is the removal of extra or undesirable soil from a location, and filling is the placement of soil where it is needed. Although this concept seems simple, conditions exist that can complicate the process. If the following logical steps are not followed, even on a limited project, there is a risk that the site will not turn out as planned:

1. *Strip sod* (remove and dispose of plant material that covers a site); see Figure 4-1. If this plant material is just moved and buried when filling, it will later decompose and settle.

2. *Strip topsoil* (remove and save **topsoil**) so that it can be respread (brought back later in the project and placed in areas where it will benefit plants).

3. *Rough grade* (cut and fill to change the grades to the elevations and forms that you desire.)

4. *Finish grade* (smooth and prepare site for planting when most of the construction has been completed and the topsoil respread.)

Determining the size of project that can be effectively graded by hand is a matter of experience and ability. It is not uncommon to move up to five cubic yards by hand. Quantities in these amounts, which would be approximately 40 wheelbarrow loads, will require significant stamina and strength. Because of the greater efficiency of heavy equipment, many grading projects are referred to grading specialists. When considering the grading portion of a project, the decision should properly match an individual's equipment and capabilities.

A project will usually move through the grading process with minimal problems, but occasionally **subgrade** soil problems must be corrected. Problem soils are identified by areas that feel spongy or that sink under foot traffic. Also look for soils that have leaf and twig debris. Drainage problems in the soil can be identified by the presence of standing water or water seeping into an excavation from any direction. During dry seasons there may be no indication of water, but mucky gray soils

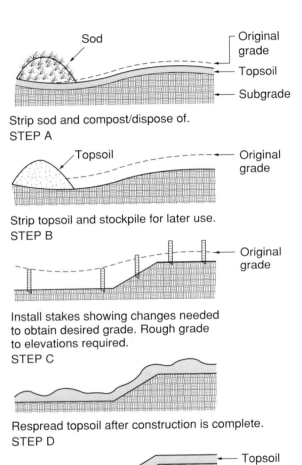

Strip sod and compost/dispose of.
STEP A

Strip topsoil and stockpile for later use.
STEP B

Install stakes showing changes needed
to obtain desired grade. Rough grade
to elevations required.
STEP C

Respread topsoil after construction is complete.
STEP D

Finish by smoothing to desired grade.
STEP E

Figure 4-1   *Steps in the grading process.*

Grading your project area

CAUTION

Verify the location of all utility lines prior to beginning construction.

Time: Varies depending on the size of the excavation; typically 2–4 hours to regrade a 10' by 10' square area 1' deep.

Level: Moderate (14 steps). Extensive digging required.

Tools Needed:

1. Project plan.

2. Several wooden stakes.

3. Two-pound sledge.

4. Marking pen.

5. Marking paint.

6. Round-nosed shovel.

7. Hand tamper.

8. Garden rake.

9. Wheelbarrow.

10. Storage area for topsoil.

Materials Needed:

Additional soil may be necessary, or disposal of excess soil may be required. Quantities of additional soil can be determined using the volume calculations described in Chapter 1.

continued

Directions:

1. Identify the general areas in which changes in elevations occur on your plan.

2. Mark the perimeter of the area in which elevation changes are planned.

3. Remove all unwanted plant material and strip sod within the perimeter of the area to be graded (Figure 4-1, step A).

4. If the topsoil in the project area is of high quality, excavate a layer of topsoil at least 6" deep and store it in a separate location to be used later (Figure 4-1, step B).

5. Within the graded area, install stakes that locate the edges and corners of proposed walls, patios, walks, structures, and other key improvements.

6. At each stake, mark how much the elevation is to be raised or lowered. Simple markings would be a plus sign for fill areas and a minus sign for cut areas, with the approximate amount noted in feet and tenths. The amount of elevation change for each location can be estimated by reviewing the plan and determining which direction the slope runs. Further adjustment of grades may be required as the project progresses, but for this step you are just trying to establish approximate, or "rough," elevations.

7. Review the stakes to determine which areas require additional soil (fill areas) and which areas require excavation (cut areas).

8. In the areas in which fill is required, use the garden rake to rough up the surface.

9. Excavate soil from the areas identified as cut, and transport it to the areas identified as fill (Figure 4-1, step C). If you place more than 6" of fill in any area, use a hand tamper to compact it before adding more fill. Excavate cautiously, particularly in areas in which you know or suspect utility lines are present. If any utilities are encountered during excavation, elevations may have to be adjusted to keep the

required amount of soil cover over the utility. Leave all stakes in place so that you can gauge whether you have cut or filled enough.

10. When the areas around all stakes have been excavated or filled, cut and fill the areas between the stakes to create even slopes from stake to stake.

11. Complete any trenching, utility, and hardscape work required for your project before completing the remaining steps.

12. If topsoil was saved, respread on areas that are to be planted (Figure 4-1, step D).

13. Once you have finished excavating, use a rake to smooth all surfaces (Figure 4-1, step E).

14. Verify that the slopes of all areas match those on the plan. Make any adjustment necessary.

that have an odor of sewage are an indicator of occasional wet conditions. If you feel that subsurface drainage problems exist, the most effective solution is to install subgrade drainage pipe in the wet area (see Installing Drainage Pipe and Inlet later in this chapter).

If unsuitable soils are present, remove these soils and replace them with a more stable material. Fill problem soil excavations in lawn areas with soil, and under pavement fill them with 1" **crushed stone**. Place these materials in layers (also termed **lifts**) up to 6" deep, until the original grade is restored. After each layer is placed, compact it with a hand tamper or a **vibratory plate compactor**. The vibratory plate compactor works like a self-propelled lawn mower. Place the equipment on the smoothed base, start the compactor, and steer. Lift the handles and twist the compactor to turn corners. By pulling back on the handles the compactor can be held in place or backed up.

Fixing poor soil problems below your project

CAUTION

Verify the location of all utility lines prior to beginning construction.

Time: 2–4 hours for repairing a 4' x 4' by 1' deep unsuitable soil area. Actual time will vary, based on the extent of poor soils.

Level: Moderate (7 steps). Digging required.

Tools needed:

1. Project plan.

2. Marking paint.

3. Round-nosed shovel.

4. Hand tamper.

5. Garden rake.

6. Wheelbarrow.

7. Optional: vibratory plate compactor.

Materials Needed:

1. Disposal area for poor soils.

2. Enough suitable fill soil or angular 1" crushed stone to fill the area excavated. Calculate the quantity after the poor soils have been excavated.

Directions:

1. Excavate the area required for your project.

2. Identify the perimeter of any area with poor soils.

3. Using the paint, mark the perimeters of the problem soils.

4. Excavate the poor soils to a depth of 12". Dispose of the poor soils in an area that will not be used for building. Excavate cautiously, particularly in areas in which you know or suspect that utility lines are present. If any utilities are encountered during excavation, elevations may have to be adjusted to keep the required amount of soil cover over them.

5. If the poor soils extend deeper than 12", continue excavating to a depth of 24". Soil problems deeper than 24" should be referred to a professional soils engineer or contractor.

6. Fill the excavated area with suitable fill soil if in a lawn area or with angular 1" crushed stone if under a paved area. Place the fill in 6" layers and compact with the hand tamper or the vibratory plate compactor. Continue placing layers and compacting until the excavated area is brought up even with the original grade (Figure 4-2).

7. Continue with your planned project.

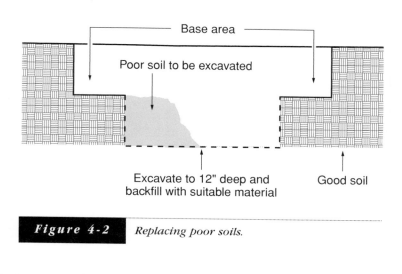

Figure 4-2 *Replacing poor soils.*

Sculpting the earth is a grading activity that can add beauty and functionality to any landscaped area, particularly areas in which the existing topography is flat. You can build a **berm**, or soil mound, to create interesting effects and to raise plant material to a higher level. Berms are also effective in creating a sense of privacy without installing a fence. Berms can take any shape or form, with interesting effects achieved from multiple berms, varying heights of berms, or creating free-form shapes.

Building a berm

CAUTION

Verify the location of all utility lines prior to beginning construction.

Time: 4–8 hours for a berm 4' wide by 8' long by 2' high.

Level: Moderate (9 steps). Digging required.

Tools Needed:

1. Plan for berm (dimensions and approximate height).

2. Marking paint.

3. Square-nosed shovel.

4. Round-nosed shovel.

5. **Rototiller**.

6. Wheelbarrow.

7. Shovel.

8. Hand tamper.

9. Rake.

10. Any required permits or approvals.

Materials Needed:

1. Enough clean topsoil to construct the berm. Use the formula presented in the section on volume calculations in Chapter 1 to determine the amount of soil required. A 10' by 10' berm 2' high in the center will require approximately 3.5 CY of soil.

Directions:

1. Mark the location of the berm.

2. Strip the sod or ground cover from the area within the perimeter of the berm, if necessary.

3. Till the existing soil within the berm perimeter.

4. Deposit soil in the berm area in layers 6" deep. Compact each layer with the hand tamper before adding another layer.

5. Continue to mound soil until the berm reaches the desired height.

6. Smooth the soil to the shape desired for the berm. The sides of the berm should have even slopes, and the soil should be tapered at the edges to meet the existing grade (Figure 4-3).

7. Verify the dimensions and elevations required by the design.

8. Adjust the grade and add or remove soil if necessary (Note: a berm compacted by hand may settle up to 10 percent, so add additional soil if the height is critical).

9. Plant or cover as desired. Provide cover in some form to reduce erosion potential.

continued

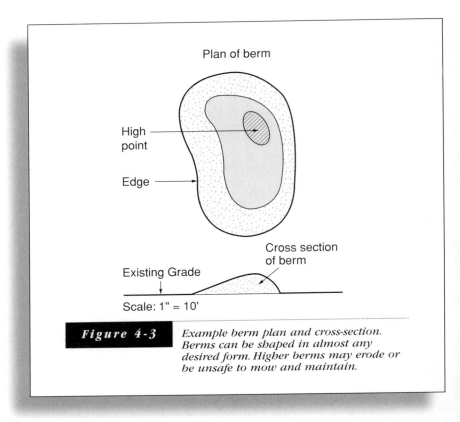

Plan of berm

High
point

Edge

Cross section
of berm

Existing Grade

Scale: 1" = 10'

Figure 4-3 *Example berm plan and cross-section.
Berms can be shaped in almost any
desired form. Higher berms may erode or
be unsafe to mow and maintain.*

EROSION PROTECTION

An unfortunate side effect of altering and sculpting land is erosion.
Erosion is the removal of soil from a site because of the action of
water and wind. Conditions that exist during the landscape-construction process, particularly site-grading, leave soils vulnerable to erosion
processes. Several factors influence erosion, including soil type, plant
cover, soil-particle size, the slope of the site, the exposure of the site,
and amount and speed of water and wind over the area. Changing
or disrupting these factors is the focus of most erosion-protection
methods. Control methods typically involve covering the soil, reducing

the amount of water and wind passing over disturbed soils, anchoring the soil with plant roots, slowing the water and wind speed to reduce their capacity to move soil particles or some combination of these methods.

The areas most susceptible to erosion include any surface that has been disturbed, especially when the slope exceeds 3 percent (approximately 4 inches of fall in 10 feet of horizontal distance). Protection methods used to reduce surface erosion include establishment of **cover crops**, **mulching**, covering sites with erosion-preventing blankets, and structural coverings on steep slopes. Each of these methods can be used as either a temporary or permanent control measure. Effective controls also include establishment of a cover crop with mulching or erosion blanket, or permanent structural controls.

Using Cover Crops to Reduce Erosion

Plants covering a site buffer the impact of falling water droplets and hold soil with their roots, providing effective erosion control. Many designs rely on vines, ground cover, and woody plant material as a permanent cover to protect slopes from erosion. Because many plants require considerable time to get established, it will be necessary for you to combine planting with temporary erosion-control measures. One temporary control is seeding a disturbed site with a cover crop. Selecting a cover crop requires knowledge of locally available crops that germinate quickly. Common selections include oats, wheat, annual ryegrasses, and buckwheat. Because of the time required for plants to germinate, planting may be used in combination with erosion-control blankets (see following) to provide more effective control. Mulching will also aid in reducing erosion.

Mulching for Erosion Protection

Providing an inert covering over a disturbed surface will reduce the impact of water droplets on soil particles. Using mulch in this way will reduce the energy with which the water strikes the soil, limiting the disturbance of soil particles and thus the potential for erosion. Mulching is a labor-intensive process with limited effectiveness on steep slopes but is appropriate for use on slopes under 3 percent. Mulch functions best when combined with seeding a cover crop under the mulch.

Mulching begins with selecting a suitable mulch material for the project site. Common choices are straw, wood chips, and shredded wood. Straw is a cost-effective temporary mulch to be used during construction or after seeding rather than as a long-term solution to erosion. Spread mulch evenly over the entire disturbed area by hand or mechanical means (a spreader), and periodically remulch during construction. Expect some minor erosion and gullying that will need to be repaired prior to respreading topsoil. If **gullying** becomes too serious a problem, consider a more intensive means of protection.

A new variation on mulching that has proven to be more effective and less labor intensive is **hydromulching**. This process mixes mulch with water and a **tackifier** (sticky substance), which is then sprayed on a disturbed area. Cover crop seed can also be mixed with the mulch, although the germination and survival rate of seed is reduced when this mixture is used. Hydromulching allows use of durable mulches and reduces the labor required to evenly spread the mulch over the site. Newer hydromulch formulas called **bonded-fiber mulch** (BFM) chemically bond the materials together and provide a durable, papier-mache like coating over an exposed area. Hydromulching is typically accomplished by feeding the materials into a trailer-mounted hopper and evenly spraying a site using a hose applicator. Because of the cost of specialized equipment, hydromulching will most likely require hiring a contractor.

Installation of Erosion Control Blankets (Erosion Blankets)

An **erosion-control blanket** (ECB or erosion blanket) consists of a layer of biodegradable materials sandwiched between two layers of lightweight netting. Blankets of this type are anchored to a disturbed surface that has been seeded, either above or below the blanket. The blanket material—wood fiber, excelsior, straw stems, or other wood by-product—degrades over one to two seasons and allows any seeded ground cover below the blankets to grow. The lightweight netting is ground up by mowers or embedded into the ground cover that eventually covers the site. Although blankets present a higher cost for erosion control than mulching, they do provide better cover and a more stable form of protection for slopes up to 4 percent (Figure 4-4).

Figure 4-4 *Placement of an erosion mat along the length of a channel. Note that the mat is placed with the long dimension running in the same direction as water will flow.*

Installing erosion blankets and cover crop

CAUTION

■ Verify the locations of all utility lines prior to beginning construction.

■ Use caution when cutting materials.

Time: 2–4 hours, depending on size of project

Level: Easy (10 steps). Digging and leveling required.

Tools Needed:

1. Square-nosed shovel.

2. Round-nosed shovel.

3. Garden rake.

4. Drop spreader.

continued

5. Sod roller.

6. Utility knife.

7. Claw hammer.

8. Rubber mallet.

9. Screwdriver.

10. Wheelbarrow.

Materials Needed:

1. Enough erosion-blanket material to cover the area plus 10 percent additional for overlap, cutting, and contingency. Calculate the amount using the area formula discussed in Chapter 1.

2. Fertilizer, 1 pound of 8-16-16 fertilizer for every 1000 SF of area to be covered by the erosion blanket.

3. Seed (preferably a fast-germinating cover or sod-forming grass). The amount of seed will vary, depending on the species of cover. Ask for advice from your seed supplier as to what species and how many pounds will be necessary for your area.

4. Landscape staples or **sod staples**. One staple for each square foot of erosion blanket required.

Directions:

1. Identify and measure the area to be covered.

2. Prepare the slope by removing any existing plant material (Figure 4-5, step A). Rake and smooth the entire seedbed below the area where the blanket is to be placed. The surface must be smooth and without ridges or valleys more than 2" in depth.

3. Using the drop spreader, place the starter fertilizer and seed over the area to be covered by the erosion blanket.

4. Excavate a 6"-deep-by-6"-wide trench along the entire length of the high side of the installation (Figure 4-5, step B).

5. Beginning at the low edge of the slope, roll out blankets in the same direction that water will run down the slope. Wherever two blankets are adjacent to one another, overlap the blankets by 12". Blankets can also be overlapped if there is excess material when covering odd-shaped areas. Overlap by placing the blanket from the higher side on top of the blanket on the lower side (Figure 4-5, step C).

6. Cut off any excess material using the utility knife.

7. Tuck the blanket into the trench at the high side. Secure every 12" with staples at the bottom of the trench (Figure 4-5, step D). Backfill and compact the trench.

8. Verify that there is good blanket-to-soil contact. Roll if necessary with the sod roller.

9. Secure the blanket, including all overlapped areas, with staples. Install sod staples or metal stakes every 12" around all edges of the blanket. Place staples in staggered rows every 18" over the interior areas of the blanket (Figure 4-5, step E). Drive the staples flush with the surface using either a claw hammer or a rubber mallet. If staples bend when installing, create a starter hole for each leg of the staple using a screwdriver.

10. Irrigate the area.

For safety, remove all stakes and staples after the ground cover is established and before the first mowing. This will reduce the chance of mowers throwing loose stakes as projectiles.

continued

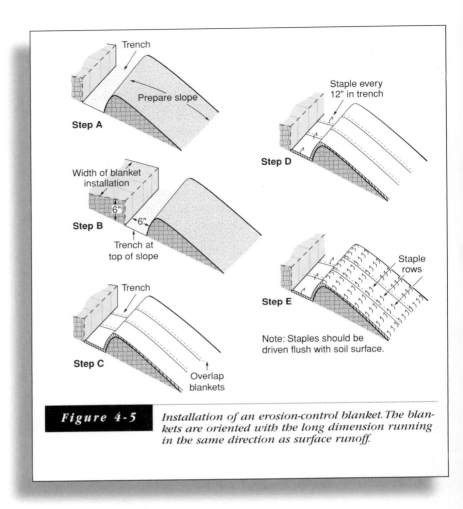

Figure 4-5 *Installation of an erosion-control blanket. The blankets are oriented with the long dimension running in the same direction as surface runoff.*

In cases in which plant material is not a reliable or practical choice for embankment or drainage-channel erosion protection, a more permanent protection can be obtained by installing **rip-rap**, or large stone. Install the rip-rap beginning at the bottom of the slope. Place a 4"–6"diameter or larger rip-rap over the exposed slope (Figure 4-6). Material choices may include rounded or angled stone. Whatever choice you select, the rip-rap must be large and heavy enough to counter the force of water washing down the slope. Extend the rip-rap 18" beyond the edges of the embankment or channel top, bottom, and sides. Rip-rap may be placed over ECB's or filter fabric to further reduce erosion.

| Figure 4-6 | *Swale protected with 4–8 inch rip-rap.* |

SITE DRAINAGE

Water that does not move properly through your site can single-handedly wreak havoc. Problems ranging from flooding and drought to freezing and thawing damage can be traced to water. Proper management of the many potential impacts of water on a site will minimize construction and long-term problems. A critical initial step in successful landscape development is the ability to drain water away from improvements on the site.

Two approaches can be used when making site-drainage improvements: drain water from the site on the surface or use underground drainage structures to store or remove water. Draining the site on the surface should be your first choice, particularly if the volume of water is small or the drainage problem intermittent. Significant and persistent problems will require that a more extensive drainage system be installed. If your drainage problem has the potential to cause significant property damage or create a hazard, seek the advice of a professional trained to design such solutions.

Surface Drainage

Surface drainage consists of grading and shaping the ground to direct runoff where desired. When preparing a site for proper surface drainage, the goal is to slope the ground away from important features that can be damaged by water, such as structures, walls, and paved areas. In addition, grading the site can eliminate low areas that will collect water and make spaces unusable after rains or irrigation. This method of removing water will prove the most cost effective and easiest to construct. It is applicable on most sites to some degree and in many sites is the only technique needed to drain the site.

Basic to the idea of surface draining a site is the principle of maintaining a low point, or a series of low points, that will keep the water level lower than structures and important site features. These low points are sometimes referred to as "**free-outs**" or "**swales**" (Figure 4-7). Free-outs are established so that if water rises to flood stage it will pass through the low point before entering a structure. This principle is easy to maintain on sites with ample topography but a challenge on sites that are flat. If your site limits the surface-drainage possibilities or makes it impossible to create low points, consider using a subsurface drainage system.

Subsurface Drainage of Surface Water

A more involved approach to draining a site is necessary when the volumes of water are great or the site itself is not conducive to surface drainage. To remedy this condition, subsurface drainage systems must be employed to maintain positive drainage. Subsurface systems collect the water on the surface and then let gravity draw it into storage or a system of pipes for removal. Although more expensive and time consuming to construct than surface drainage, subsurface systems are effective in draining a site when properly maintained.

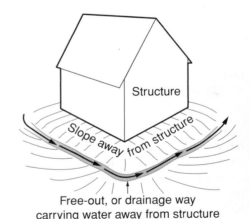

Free-out, or drainage way carrying water away from structure

Figure 4-7 Surface drainage free-out. Runoff from the hill and water directed away from the structure is picked up in the drainage swale and then diverted away from improvements.

An initial step in implementing subsurface drainage is to store excess water below grade in a **French drain**. French drains are trenches filled with stone and covered with a thin layer of soil and ground cover. Water seeps into these trenches and is stored in the open space between each stone until it can seep into the surrounding soil. French drains are considered a closed drainage system, requiring no outlet point. Best used to address standing water after a storm or irrigation, French drains are most effective when installed in low areas that collect water and in which compacted soils slow water percolation. Drawbacks to this system are that French drains are not suitable for persistent heavy flooding and can become saturated during wet seasons, while lawn above the trench may dry sooner than surrounding turf during dry conditions.

Installing a French drain

CAUTION

- Verify the locations of all utility lines prior to beginning construction.
- Use caution when cutting materials.

Time: 4–8 hours for a 10' long drain.

Level: Moderate (8 steps). Digging required.

Tools Needed:

1. Marking paint.

2. Round-nosed shovel.

3. D-handled, flat spade.

4. Pick.

continued

5. Wheelbarrow.

6. 25' tape measure.

7. Utility knife.

Materials Needed:

1. Enough 1"–2"-diameter rounded stone (**washed river rock**) to fill a 12"-wide, 30"-deep trench the length of your French drain. Use the volume formulas presented in Chapter 1 to calculate the required quantity.

2. A length of 36"-wide **landscape fabric** twice as long as the length of your French drain plus four feet.

3. Disposal area for excess soil.

Directions:

1. Locate the area of your lawn that does not drain. Using the marking paint, mark the location of the trench. To be effective, the majority of the trench should be located directly under the area that floods.

2. Excavate a trench 12" wide and 36" deep (30" for the drain and 6" for the cover material) along the alignment of the trench. In turf areas the sod may be removed and set aside for reuse (see Project: Removing sod). Save approximately 10 percent of the soil excavated from the trench for cover (Figure 4-8).

3. Using the utility knife, cut two lengths of 36"-wide landscape fabric 2' longer than the trench length.

4. Place the first piece of landscape fabric in the trench against one side. Fold the top over the outside edge of the trench. It is not necessary to cover the entire bottom of the trench.

5. Repeat step 4 of this process with the other piece of landscape fabric on the opposite side of the trench.

6. Fill the trench to within 6" of the top of the trench with 1"–2" diameter stone. When filling the trench, take care that the landscape fabric is not pulled out of position.

7. Fold the remaining landscape fabric over the top of the river rock.

8. Backfill the trench with the soil set aside earlier and compact it. If sod was saved, replace over the fill.

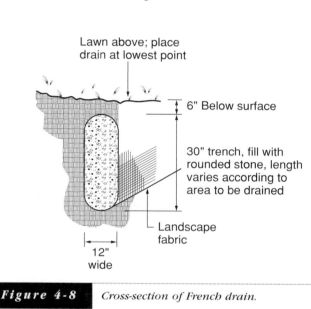

Lawn above; place drain at lowest point

6" Below surface

30" trench, fill with rounded stone, length varies according to area to be drained

Landscape fabric

12" wide

| **Figure 4-8** | *Cross-section of French drain.* |

When volumes of water are great enough that they cannot be stored below grade in a French drain, they will have to be collected and carried away using a subsurface drainage system. Subsurface drainage systems can be designed to drain away both surface water and subsurface water. Surface water is drained away through the use of **inlets** to collect water and funnel it into a **nonperforated plastic pipe** to be carried to an outlet point. Subsurface drainage systems are used to intercept the **water table** (subsurface water that is moving up toward the surface) and water infiltrating downward into a **perforated plastic pipe** and direct it to an outlet point.

With both surface and subsurface drainage systems, consider using erosion control at the outlet points. At these locations the volume of water is typically heavy and can be fast moving. These conditions are the formula for erosion. To reduce the potential for erosion, place rip-rap around the outlet point of any drainage swale or drainage pipe.

The installation of a subsurface drainage system to drain surface areas begins with the flagging of inlet points and an outlet point. Inlets should be located at low points in paved areas or lawns, near roof downspouts, and in any location in which significant runoff is expected. If necessary, you can install multiple inlet points, using a simple network of plastic drainage pipe connecting several inlets and directed to the outlet point.

Installing a subsurface drainage system with an inlet

Alternatives to this project: If the area is too large to drain with a single inlet, a multiple-inlet system may be necessary. See Appendix D, Figure A-11 for details on a multiple-inlet system.

CAUTION

■ Verify the locations of all utility lines before beginning construction.

■ Verify that it is legal to empty water from the outlet of your plastic drainage pipe at the location planned.

■ Use caution when cutting materials.

Time: 4–5 hours for every 10 linear feet (LF) of pipe.

Level: Moderate (11 steps). Extensive digging required.

Tools Needed:

1. Surveyor's flags.

2. Carpenter's level.

3. Round-nosed shovel.

4. D-handled, flat spades.

5. Garden rake.

6. Pick.

7. 25' tape measure.

8. Utility knife.

Materials Needed:

1. Duct tape.

2. Enough 4" nonperforated plastic drainage pipe to run the length of your drainage system, plus 10'. This pipe comes in rolls of up to 100' and in 10' sections.

3. Premanufactured 4" 90-degree elbow, one for every inlet.

4. Fiberglass inlet grate for 4" nonperforated plastic drainage pipe, one for every inlet.

Note: Multiple inlets may require additional pipe, grates, and T fittings. See Appendix D for illustration.

Directions:

1. Flag an inlet point (the low point of an area you want to drain) and an outlet point (a lower point where you can empty the water you have collected; this may be a waterway, storm drain, curb, or a lower portion of your yard). The outlet point must be lower than the bottom of your plastic drainage pipe to maintain a slope that drains. Your route should not intersect any utility lines that are at the same depth as the drainage system.

continued

2. Beginning at the inlet point, excavate a trench that is 8"–12" wide and at least 12" deep and gradually sloping downward from the inlet point to the outlet point (Figure 4-9).

3. Beginning at the inlet point of the trench, lay 4"-diameter non-perforated plastic drainage pipe along the bottom of the entire trench. A continuous roll of plastic drainage pipe may run the entire length, but shorter pieces of plastic drainage pipe will need to be joined by sliding the enlarged end over the smaller end of the previous piece.

4. Place a carpenter's level on top of the plastic drainage pipe every 5' to verify the downhill slope of the pipe. If the slope does not run downhill, remove the pipe and excavate deeper.

Figure 4-9   *Excavating a trench and placing a plastic drainage tile.*

5. At the location at which the inlet is planned, cut the end of the plastic drainage pipe square using a utility knife and insert a premanufactured elbow. Slide the elbow over the

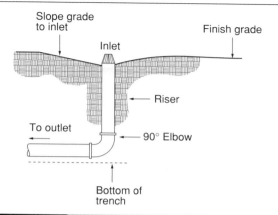

Slope grade to inlet

Finish grade

Inlet

Riser

To outlet

90° Elbow

Bottom of trench

Figure 4-10 *Cross-section of plastic-drainage-pipe riser. The inlet captures water runoff and directs it into the drainage pipe.*

end of the plastic drainage pipe and snap it into place. Aim the open end of the elbow toward the surface (Figure 4-10).

6. In the opening of the elbow, insert a vertical plastic drainage pipe (riser) running from the elbow to approximately 12" above the surface (Figure 4-11).

7. To create a more secure installation, wrap duct tape around any joints to reduce the possibility that connections might pull apart.

Figure 4-11 *Plastic-drainage-pipe riser before being trimmed for grate installation.*

continued

8. Backfill the trench to the surface, lightly compacting the backfill after every 6" layer.

9. Around the inlet location, regrade the surface to slope to the inlet (now marked by the riser you left extended out of the ground).

10. When the grade has been adjusted to drain to the inlet, cut the riser flush with the surrounding elevation and insert a pre-manufactured fiberglass inlet grate into the end of the riser.

11. Install rip-rap around the inlet to reduce erosion and if necessary.

Subsurface Interception Of Water

In some areas, water problems may be caused by water moving up from subsurface areas rather than by surface water failing to drain away. An area with excess subsurface water can be identified by wet, spongy lawns. Also, water standing on the surface, even when there has been no rain or irrigation, is an indication of a subsurface water problem.

If you need to remove subsurface water from a site, you can adapt a drainage system similar to the subsurface system with inlets to solve the problem. Build an underground system similar to the inlet system, but omit the surface inlet points and use perforated socked plastic drainage pipe. Place a fitted cap over the high end and outlet in a similar manner as the inlet system. The perforated pipe allows water to enter the system and be drained away before it reaches the surface.

Chapter 5

LANDSCAPE RETAINING WALLS

On sites that have a significant amount of slope, normal use of open areas for lawns, gardens, play areas, or circulation routes may be restricted because of the steep grade. **Retaining walls** can create level areas that will accommodate these activities. Whether designed as a single element or several walls grouped together in **terraces**, the immediate impact of this vertical element will create a significant change in a landscape. Add stairs to your design, and what was once an unusable hillside is now a terraced garden. Walls can also be arranged to create planters and raised beds or kept low to define the edges of gardens.

Retaining walls can also reduce erosion by eliminating steep, erodable slopes, but are a costly way to address the problem. In addition to their dominant appearance, walls are an expensive and time-consuming landscape element to construct. If erosion control is your primary purpose for erecting a wall, consider all possible solutions. Slopes that are graded and planted can perform many similar functions at a much lower cost.

Whatever the function of your wall or walls, contemporary construction materials and techniques make walls a versatile and effective landscape element. Walls of almost any form, color, and material can be blended with other landscape elements to accomplish many design goals. Many choices of material are available for building landscape retaining walls. Walls that are within the price and skill range of the typical homeowner include **wood landscape timber** walls, **segmental wall block (precast concrete)** construction, and **dry-laid-stone** wall construction. Other types of walls better suited for installation by a contractor are retaining walls of **cast-in-place concrete**, **mortared veneer** materials (concrete walls with stone laid in front for esthetic enhancement), and **gabions** (stacked wire cages filled with stone).

Although the homeowner may have the ability to build walls using various materials, the height of a retaining wall built with any material should be limited. Walls over 30" high require special anchoring techniques reserved for the experience and capabilities of a landscape contractor.

PLANNING FOR WALL CONSTRUCTION

Planning a retaining wall project requires considering all factors that will contribute to a successful installation. Although material choice is important, it is equally important to consider where the wall will be placed, how to construct a sound base below the wall, and how to preserve the installation by using proper height, drainage, and stabilization. It will also be beneficial to know how your wall will end and how to terrace a site when one tall wall is impractical. You should also check local regulations to identify any restrictions on wall locations, materials, and heights that may apply to your project.

Wall Layout

Before beginning construction, identify the location of the front of the wall. Place stakes or paint the ground along the alignment to identify potential starting points and any problems that may be encountered. Consider avoiding soils that are unstable or areas in which significant amounts of water accumulate or pass through. Verify any locations where corners or stairs may be required. When laying out the wall alignment, check slopes to assure that a wall of 30" height or less will perform the functions needed. If a higher wall is needed, consider building a terraced installation (see following).

Excavation and Base Preparation

Excavation will be required to prepare the base for a wall, and may be extensive at locations in which the grade is not consistently level. At the low point of the alignment, begin excavating sod and topsoil from an area in front of and behind the area where the wall will be placed. The excavation for the base must be deep, because the base material and the first **course** (layer of stones or other material) in any wall will be below grade when the project is finished. When excavating, any unstable materials you encounter should be removed and replaced with **granular backfill** (see Project: Fixing poor soil problems, Chapter 4).

If possible, the bottom of this trench should be level from end to end. In sloped areas the depth of this level trench may exceed the thickness of the wall material being used. In these cases the wall should be stepped up by raising the bottom of the trench. When walls encounter embankments, this stepping may occur rapidly, but to maintain stability, always step the wall up one course at a time. When the wall steps, the bottom course of the wall should still be buried below grade.

Fill this trench with **free-draining angular crushed stone** for stability, and then level and compact it using a hand tamper or a vibratory plate compactor. A 1/2" layer of **fine granular material** (sometimes called stone dust, decomposed granite, or 3/8's minus) will simplify the leveling of the first course of wall material.

Installing the base for a retaining wall

CAUTION

- Verify the location of all utility lines prior to construction.
- Follow the manufacturer's instructions when using equipment.

Time: Varies, depending on wall length and amount of soil to excavate. Approximately 15–30 minutes for each linear foot of wall being built.

Level: Moderate (8 steps). Heavy lifting required.

continued

Tools Needed:

1. Marking paint

2. 25' measuring tape.

3. Plan for your wall.

4 Round-nosed shovel.

5 Square-nosed shovel.

6. Carpenter's level.

7. Wheelbarrow.

8. Garden rake.

9. Hand tamper.

10. Optional: Vibratory plate compactor.

Materials Needed:

1. Enough free-draining angular 1" crushed stone (also termed rubble or washed class 2 aggregate) to fill a trench 6" deep, 30" wide, and the length of your wall. Use the volume formulas presented in Chapter 1 to calculate quantities.

2. Enough finely crushed stone (stone dust, or stone that will pass through a 3/8" sieve) to place a 1/2" deep layer over the entire base; approximately 1 cubic foot (CF) for every 10 LF of 30" wide trench.

Directions:

1. Lay out the alignment of the front of the wall and mark the location with marking paint.

2. Mark lines 6" in front of and 12" behind the entire wall alignment (Figure 5-1). Ensure that the back mark is 12" from the back side of the wall material.

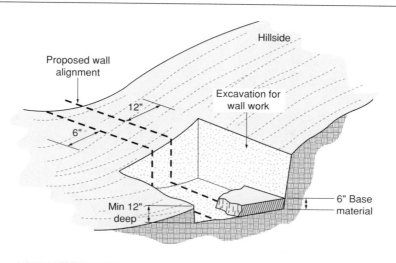

Hillside

Proposed wall
alignment

Excavation for
wall work

12"

6"

6" Base
material

Min 12"
deep

Figure 5-1 *Preparing an excavation for retaining walls.*

3. Remove the sod or ground cover between the front and back mark.

4. Beginning at the low point of the wall alignment, excavate a 12" deep trench along the entire area where sod or ground cover was removed. Use the carpenter's level to verify that the bottom of the trench has remained level. The depth of the trench is to allow for 6" of base material and the bottom course of the wall material (Figure 5-2). Soil excavated from this trench can be used for backfill behind the wall. If the level trench reaches a depth of more than twice the thickness of the wall material (typically 18"), you should step both the wall base and the wall up one course.

5. Place free-draining angular 1" crushed stone in the bottom of this trench to a depth of 6".

6. Using the rake, level the base material. Use the carpenter's level to verify the surface of the base material is still level. Adjust if necessary by adding or removing base material.

continued

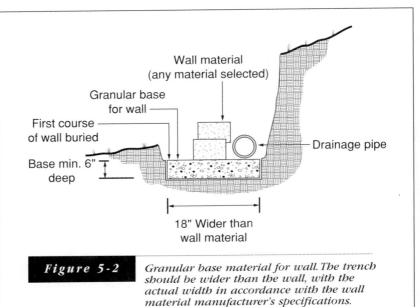

Wall material
(any material selected)

Granular base
for wall

First course
of wall buried

Drainage pipe

Base min. 6"
deep

|←——— 18" Wider than ———→|
wall material

Figure 5-2 *Granular base material for wall. The trench should be wider than the wall, with the actual width in accordance with the wall material manufacturer's specifications.*

7. Compact the base material using the hand tamper or a vibratory plate compactor.

8. Add a 1/2" thick layer of finely crushed stone (that which passes through a 3/8" sieve) on top of the 1" stone to make leveling of the first course of wall material easier. The base is now prepared for placing wall material.

Beginning Walls

After the base installation is complete, survey the site to determine the lowest point. Placement of wall material should begin at the low point of the wall; beginning the bottom course at the low point allows all higher courses to be stacked on top of the lower one. However, if the entire base trench is level, construction can begin at either end. If you begin the wall at a higher point, it will be difficult to fit wall material below an upper course already in place. If your wall must end evenly at a structure or other permanent improvement that is not at the low point, carefully measure from that ending point to the low point. This measurement will allow you to slightly adjust the placement of the beginning of the wall and reduce the chances of having to cut materials to fit when you reach the end.

Drainage Behind Walls

Wall failure can often be traced to a buildup of **hydrostatic** (water) pressure behind the wall. The forces exerted by water behind a wall are strong enough to collapse even a reinforced installation. You can release water from behind a wall by drilling holes through the base course of the wall every 6 to 8 feet (**weep holes**), or by allowing water to pass through naturally open wall joints (narrow openings between wall materials).

Although weep holes and wall joints are beneficial, supplemental drainage is recommended for any type of wall that exceeds 2' in height. You can accomplish this by installing perforated plastic drainage pipe placed behind the base course of the wall. A 4" plastic perforated **socked pipe** (the sock is a filtering fabric wrapped around the tile that reduces the amount of silt that enters the pipe) placed in this location along the entire length of the wall will aid in the removal of excess water (Figure 5-3). The drainage pipe behind the wall must be sloped to the low end of the wall or inserted through the wall (Figure 5-4) if the lowest point is not at an end. In order to work properly, free-draining angular backfill should be placed in a 12" wide zone above the drainage pipe to allow water to seep down to this pipe. This free-draining backfill should extend from the drainage pipe to just below the surface behind the wall (Figure 5-5).

Some consultants recommend placing landscape fabric behind the compacted fill material to reduce the movement of soil through the wall openings. In most wall construction projects this approach is *not* recommended, because it can increase the hydrostatic pressure behind the wall and counter the benefits gained from using drain tile and free-draining backfill.

Figure 5-3 *Placement of socked drain tile behind retaining walls.*

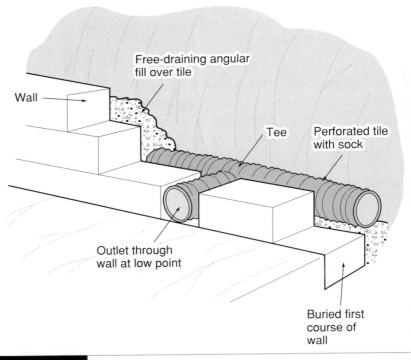

Free-draining angular fill over tile

Wall

Tee

Perforated tile with sock

Outlet through wall at low point

Buried first course of wall

Figure 5-4 | *Placement of drain tile behind a retaining wall.*

Compaction Behind Walls

The remainder of the backfill behind the angular fill may be original or imported soil. Backfill material should be placed in lifts, or layers, of no more than 6" before compaction. Select compaction methods carefully to reduce the chance of wall collapse. For a distance of 5' behind the wall, compaction should be performed by hand or with a vibratory plate compactor (Figure 5-5). Avoid overcompacting immediately behind the wall to prevent displacement of wall material. You may find it useful to backfill after laying each wall course, placing first the angular fill and then any soil backfill.

Wall Heights And Stabilization

As a wall increases in height, the possibility of its failing increases. Drawn downward by gravity and pushed forward by water pressure, the forces that can topple a wall must be countered by measures that

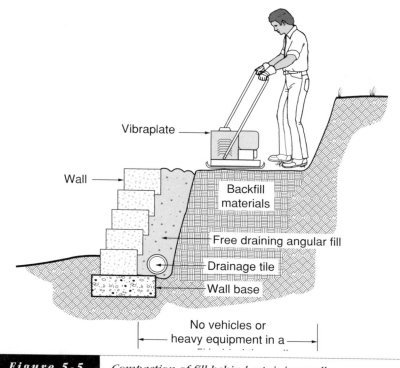

Vibraplate

Wall

Backfill
materials

Free draining angular fill

Drainage tile

Wall base

No vehicles or
heavy equipment in a

Figure 5-5   *Compaction of fill behind retaining walls.*

will stabilize it. For a short wall, burying the first course and leaning
the wall backwards with each course (**batter**) will, in concert with
drainage behind the wall, work to improve stability. For higher walls,
contractors use additional wall-stabilization techniques, such as
deadmen, geogrid, and stabilizing anchors. Deadmen are horizontal
wall members anchored in the hillside behind the wall. Geogrid is a
woven fabric that performs the same function.

Although it may seem that you are burying expensive material, placing
the first course of the wall below grade provides the first measure of
stability. The soil in front of this buried first course prevents pressure
behind the wall from pushing the bottom out. Bury, at a minimum, the
first course of your wall, and for higher walls, it is advisable to bury
additional courses. For all wall construction, excavate the base trench
low enough to place at least one full course of material below the
finish grade on the front side of the wall.

Batter is the backward leaning, or stepping, of a wall. Batter can be built into a wall using a **lean-back batter**, or tilting the base course slightly backward (approximately 1/4" fall from front to back). This will cause all subsequent courses to lean the same direction when placed. Batter can also be built into the wall by setting the front of each subsequent course back from the front of the previous course a small amount. This process, called **step-back batter**, also helps stabilize the wall but keeps the materials level. Standards for step-back batter range from 1" to 2" for each foot of wall height. Although some interlocking wall materials are designed to be set vertically using pins to anchor courses, most walls incorporate one of these two types of batter. Some wall materials have lips on the bottom of the back side that incorporate a step-back batter into the wall material automatically as it is stacked (Figures 5-6 and 5-12).

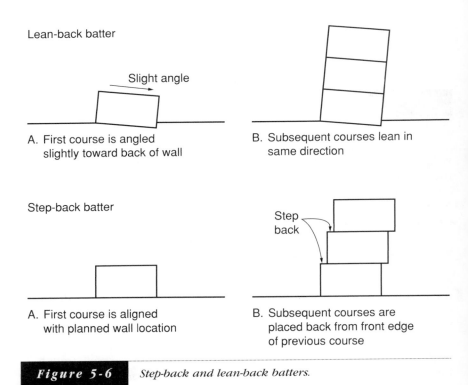

Lean-back batter

Slight angle

A. First course is angled slightly toward back of wall

B. Subsequent courses lean in same direction

Step-back batter

A. First course is aligned with planned wall location

Step back

B. Subsequent courses are placed back from front edge of previous course

Figure 5-6   *Step-back and lean-back batters.*

For a small residential project, you may want to experiment with both methods before completing an entire wall installation. Although both forms of battering walls provide effective stability, some designers consider the step-back method more attractive. You may also find the step-back method easier to install, particularly if your project has stairs or curves.

Terracing

One effective technique for reducing the possibility of wall failure is to limit the height of a wall. If you need to build on a steep slope, either terrace the site or consult an engineer. Terracing is the construction of a series of short walls, each stepped back a calculated distance from the previous one. By using terracing, what would have been a 6' wall can be constructed as two 3' walls without extensive engineering. Terracing will be of limited usefulness in areas in which there is extensive fill that could settle or in locations that are not large enough to provide proper spacing between walls (Figure 5-7).

Figure 5-7 *Terracing of segmental precast concrete retaining walls for stability. (Courtesy of Gary Pribyl, King's Materials)*

Terracing should be constructed beginning with the lower wall first, and then grading an area behind the wall and constructing the next higher wall. Generally, the spacing between the walls should be no less than twice the height of the taller wall, but if the wall is built on fill or has heavy traffic above, consult a design professional to determine the proper method for stabilization. The grade between the walls should slope 6" or more towards the lower wall.

Ending Walls

Ending retaining walls without an abrupt stop requires some method of transitioning the grade around or in front of the wall. Methods used range from **tapering** the grade in front of a wall (Figure 5-8), to stair-stepping the top of the wall to meet the grade. Your choice of a method depends on whether the grade behind the wall falls or moves upward behind the wall.

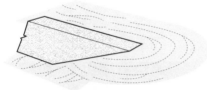

A. Tapering grade by sloping fill in front of wall to match the grade between the front and back of wall

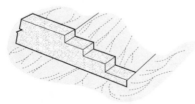

B. Stepping wall down by reducing the height of the wall over a short distance

Figure 5-8   *Ending retaining walls.*

The easiest method for dealing with the grade at the end of a wall is to construct the wall, backfill behind the wall, and then fill in front of the wall so the slope gradually tapers down to meet the grade in front of the wall. Place fill carefully so that the wall is not disturbed.

When the grade behind the wall drops, the best way to end the wall is to gradually step the top of the wall down. Execute this transition by ending the top course and extending lower courses a few blocks beyond that ending point. Continue stepping down until the wall matches grade. This may require special caps for pre-cast wall blocks to hide the exposed interior of the stepped wall. Cut and install cap stones if necessary.

BUILDING RETAINING WALLS

Your choice of wall materials should be based on several factors, including cost, appearance, ease of installation, and the durability of the material. From the entire range of wall-material choices, three typically dominate use in residential settings: wood landscape timbers, segmental wall units (precast concrete blocks), and stone. Each of these choices have pros and cons from a design-and-engineering standpoint, but all are products that are available to the homeowner and can be used by anyone with average skills in construction.

Wood Landscape Timbers

When a natural look is desired, wood landscape timbers are often used to create walls of all sizes and shapes. Manufactured to consistent dimensions and lengths, landscape timbers are a cost-effective way to assemble a retaining wall. Landscape timbers are typically sold in 6"x 6" or 8"x 8" dimensions, vary in length from 6' to 16', and have squared edges. Timbers are lightweight (when compared to other wall materials) and are easily cut and drilled.

Landscape timbers are treated so that they resist decay when placed in contact with soil. If you cut a timber before installing it, you will need to treat the cut ends in order to make them resistant to rot and insect damage. Despite this treatment, the life expectancy of all wood products is limited. Because the exact length of service for wood installed in a specific location cannot be accurately predicted, you should consider other materials that will give you a more permanent installation. In addition, some methods used to treat timbers—most recently CCA, (which leaves the wood a pale green color) is scheduled to be phased out of production due to environmental concerns. Alternatives that are safer do exist, but their durability has yet to be verified.

Do not confuse landscape timbers with the small edging and planter timbers or railroad ties sold in lumber centers. The much smaller edging timber has rounded edges and is not suitable for walls. Also, ties are treated with chemicals that are unsuitable for handling and installing plants nearby, and too irregular for simple construction. Neither of these products will produce a suitable retaining wall.

Building a timber retaining wall

1. Follow the manufacturer's instructions when using equipment.

2. Use caution when cutting and installing wall materials.

3. Obtain assistance when lifting wall materials.

4. Do not construct any wall over 30" high without assistance from a qualified designer or contractor.

Time: Varies, depending on wall length and the amount of soil you will need to excavate; approximately 1–2 hours for each linear foot of wall being built.

Level: Challenging (15 steps). Heavy lifting and equipment operation required.

Note: You should complete the base for your wall prior to installation. See the Project: Installing The Base for a Retaining Wall presented earlier.

Tools Needed:

1. Round-nosed shovel.

2. Square-nosed shovel.

3. Wheelbarrow.

4. Torpedo level.

5. Carpenter's level.

6. Utility knife.

7. Screwdriver.

8. Two-pound sledge.

9. Heavy-duty electric drill.

10. Power cords run to a GFCI outlet.

11. 3/8" × 12" long wood drill bit (auger type preferred).

12. Bow saw.

13. Paint brush.

14. Broom.

15. Hand tamper.

16. Optional: chainsaw.

17. Optional: vibratory plate compactor.

Materials Needed:

1. Enough timbers to construct the wall. To calculate the number of 8' long by 6" × 6" timbers required, measure the square footage of the face of the wall (height of wall plus 1 foot, times length) and divide by 4. Order a couple more timbers than you think you will need, just to ensure that you have enough.

2. 12" × 3/8" spikes. To calculate the number of spikes, multiply the number of timbers purchased times 4.

3. Enough socked 4" perforated plastic drain tile to run the entire length of the wall, plus 5 feet, to ensure that you have enough. Purchase a T fitting if your drain runs through the face of the wall.

4. Free-draining angular 1" crushed stone. To calculate the number of cubic yards needed, divide the square footage of the face of the wall by 27 (there are 27 cubic feet in a cubic yard).

5. 1 gallon of wood preservative rated for below-grade applications.

continued

Directions:

Placement of the first course:

1. Beginning at the lowest level of the base, place a timber on base material aligned with the front of the wall (Figure 5-9, step A). Beginning at a structure or corner is best if it is also the lowest point.

2. Using a level, check the timber to ensure that it is level from end to end (Figure 5-9, step B). If the timber is not level, add or remove base material from under one end of material.

3. If using step-back batter, use a level to check that the timber is level from front to back (Figure 5-9, step C). If the level is not correct, tilt the timber backward and remove or add base material as necessary. If you are using lean-back batter, position the timber with 1/4" slope from front to back.

4. Recheck that the timber is level both side to side and front to back after adjustments have been made (Figure 5-9, step D). Adjust again if necessary.

5. Place the next timber with the end abutting the first timber. Place your hand on the joint between the timbers to verify that the top and front face of the second timber are flush with the previous one (Figure 5-9, step E). Adjust the second timber if it is too high or too low, and then proceed to level the second timber in the same way you leveled the first one.

6. Continue to lay timbers end to end along the entire length of the first course. All timbers should be level and the tops flush. If a timber extends beyond where you want to end the wall or turn a corner, trim the tie or timber to the correct length, and then liberally brush wood preservative on the cut ends before installing.

If the site requires that the wall steps up:

1. If your wall is to step up, complete the lower course of wall to the point at which the step is to occur.

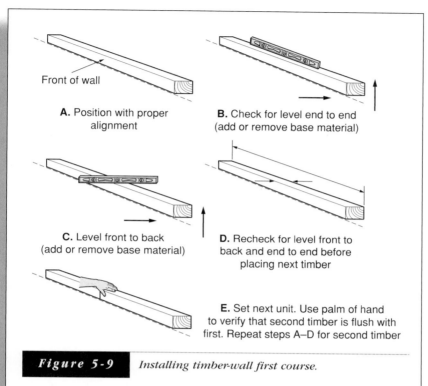

Front of wall

A. Position with proper alignment

B. Check for level end to end (add or remove base material)

C. Level front to back (add or remove base material)

D. Recheck for level front to back and end to end before placing next timber

E. Set next unit. Use palm of hand to verify that second timber is flush with first. Repeat steps A–D for second timber

Figure 5-9 *Installing timber-wall first course.*

2. Adjust the granular base for the next course so that it is level with the top of the lower timber.

3. Set a second course of timbers straddling the base and the lower course. Do not interrupt the staggered pattern when the wall steps up. If necessary, trim a timber to maintain a consistent pattern and treat the cut ends with wood preservative (Figure 5-10).

Adding additional courses:

1. Set a second course of timbers on top of the bottom course. If you are using a step-back batter, the front of the second course should step back approximately 1" from the front of the bottom course. If you are using a lean-back

continued

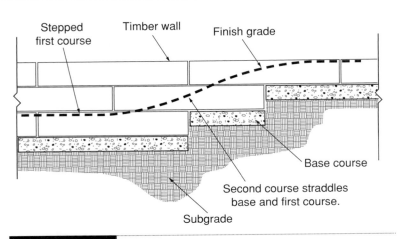

Stepped first course

Timber wall

Finish grade

Base course

Second course straddles base and first course.

Subgrade

Figure 5-10 *Stepping wall up a hill. When two courses become buried, the wall should be stepped up. Step the base and first course up one level.*

batter, the timbers should lean at the same angle as the course below.

2. The joints between timbers on each course should be staggered (offset) from the joints of the course below (Figure 5-11). If necessary, trim a timber so that you can stagger the joints. Treat the cut ends with wood preservative before installing.

3. At the end of each timber in the second course and 12" on either side of any joint in the course below, drill a 3/8" diameter hole completely through the timber. Using the 2-pound sledge, drive a 12" spike through the hole and into the timber below. Stabilize the wall to prevent it from collapsing while you drive the spikes (Figure 5-11).

4. When the second course has been completed, place a 4" diameter perforated drain tile behind the wall. Extend the tile along the entire length of the wall and run it around to the front of the wall at the low end or through the wall at the low point (Figure 5-4).

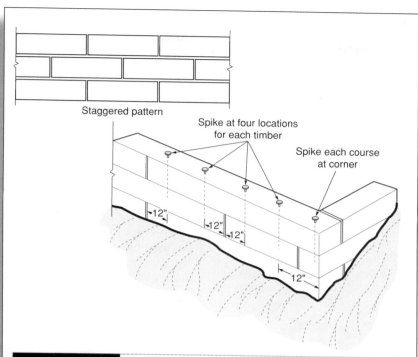

Staggered pattern

Spike at four locations for each timber

Spike each course at corner

12" 12" 12" 12"

Figure 5-11 *Spiking pattern for staggered-pattern timber walls.*

5. When the second course has been completed, set each subsequent course in a similar manner. Remember to batter the course, stagger the joints, and spike at the specified location for each timber.

6. After every course, backfill in the 12" zone immediately behind the wall with a layer of angular 1" crushed stone. The remainder of the backfill area can be composed of soil excavated from the base (Figure 5-10).

Segmental Wall Blocks

Many manufacturers now produce segmental, precast concrete blocks used for constructing walls. Similar in surface area to concrete blocks, these wall blocks come in a variety of forms, shapes, and installation methods. Segmental blocks create the appearance of a rough-surfaced stone wall, and, considering the variety of colors available, provide an attractive and consistent wall surface. Wall blocks can be heavy, ranging from 30 to 80 lbs per block, and often require the efforts of two people to lift and install. Installing them requires careful layout and base-course preparation and the stacking of subsequent courses. Walls with angled corners can require difficult cutting and spacing, so you may want to consider hiring a professional contractor to install them. Costs for segmental units are in the midrange to slightly high range when compared to other wall materials.

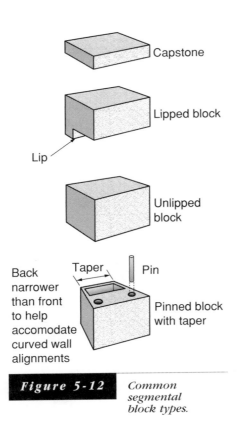

Figure 5-12 *Common segmental block types.*

Two common types of segmental blocks are unlipped blocks, which simply stack one block on top of another, and **lipped blocks**, which have a protrusion on the bottom of the block to tie the layers together (Figure 5-12). Most unlipped and lipped blocks are designed for installation using a step-back batter. Pinned and anchored forms of blocks are used by contractors for installing tall walls. Special **capstones** are also manufactured for wall-block systems that have open cores.

Dry-Laid Stone

Drylaid stone is considered a classic wall material. As the name suggests, this natural wall material is installed without mortar between the stones. If your project includes a stone wall, a sound design concept is to expand the use of stone to paving and edging. This additional use of stone will improve the unity of a design.

A stone wall can be constructed of stones of regular or random thickness and with a variety of stone types. Costs will vary but are typically high due to availability and quarrying costs, high demand in construction, and increased labor in preparation and installation. Stone can vary in size from small field rubble to large pieces that require more than one person to lift.

Note: Walls built of this material are generally stable if properly built, but walls built too high may deteriorate or collapse.

Building a segmental block (precast concrete) retaining wall

CAUTION

- Follow manufacturer's instructions when using equipment.
- Use caution when cutting and installing wall materials.
- Obtain assistance when lifting wall materials.
- Do not construct any wall over 30" high without assistance from a qualified designer or contractor.

Time: Varies, depending on wall length and the amount of soil it will be necessary to excavate; approximately 1–2 hours for each linear foot of wall to be built.

Level: Challenging (24 steps). Heavy lifting required.

Note: Complete the base for your wall prior to installation. See the Project: Installing The Base for a Retaining Wall presented earlier.

Tools Needed:

1. Round-nosed shovel.

2. Square-nosed shovel.

3. Wheelbarrow.

4. Torpedo level.

continued

5. Carpenter's level.

6. Claw hammer.

7. Rubber mallet.

8. Utility knife.

9. Screwdriver.

10. Hand tamper.

11. Optional: caulking gun.

12. Optional: vibratory plate compactor.

Materials Needed:

1. Enough block to construct the wall. To calculate the amount of block required, measure the square footage of the face of the wall (height plus 1 foot times length). Many types and sizes of blocks are available, so contact your wall material supplier to find out how many square feet of wall surface each block will cover. Divide the square footage by this number to determine how many blocks are required. Order a few extra to ensure that you will have enough.

2. Enough capstone to cover the top of your wall. If you know the length of your wall, your wall-material supplier can tell you how many will be needed.

3. Enough socked 4" perforated plastic drain tile to run the entire length of the wall plus 5 feet. Order a T fitting if your drain tile passes through the wall.

4. Free-draining angular 1" crushed stone. To calculate the cubic yards needed, divide the square footage of the face of the wall by 27.

5. Optional: 1 tube of construction adhesive for every 10 cap-stones installed.

Directions:

Placement of the first course:

1. If using a lipped block, break the lip off each block to be placed in the first course using the claw hammer. Removing the lip will make leveling easier.

2. Begin at the lowest point along the base. If the base is perfectly level, begin at one end of the wall installation.

3. Place the first block aligned with the front of the wall (Figure 5-13, step A).

4. With a torpedo level, check to ensure that the block is level from side to side (Figure 5-13, step B).

5. If the block is not level, adjustments may be made by tipping the block backward and using the fingertips to brush away base material if the block is high or by sprinkling handfuls of fine granular material under the the block if it is low (Figure 5-13, step C). Minor changes may also be made by twisting the material or by tapping lightly with a rubber mallet. When adjusting for level, use caution not to disturb any adjacent blocks that have already been set.

6. If using step-back batter, follow the same process to level the block front to back (Figure 5-13, step D, and Figure 5-14). Adjust if necessary. If using lean-back batter, position the block with 1/4" slope from front to back.

7. Recheck the block in both directions again after all adjustments have been made (Figure 5-13, step E). Adjust again if necessary.

8. Place the second block next to the first and verify that the edges are in contact and the tops are flush (Figure 5-13, step F). Level the second block using the same process used for the first block.

9. Repeat this process for each block along the entire length of the first course. Check the overall level periodically using a carpenter's level long enough to cover three blocks, and

continued

verify that all blocks are in contact with the blocks on either side.

If the site requires the wall to step up:

1. In locations in which the wall is to step up, complete the lower course of wall to the point at which the step will occur.

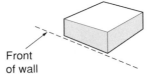

Front of wall

A. Position with proper alignment

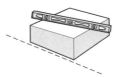

B. Check level side to side, adjust if necessary

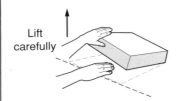

Lift carefully

C. Adjust by lifting and adding or scraping away material. Block can also be adjusted small amounts by twisting unit or striking it with a rubber mallet

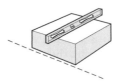

D. Level front to back; adjust if necessary

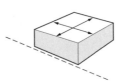

E. Recheck level side to side and front to back before placing next unit

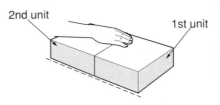

2nd unit   1st unit

F. Set next unit. Use palm of hand to verify and unit is flush with first unit. Repeat steps A–E for second unit. Use long level over several units to check side to side level

Figure 5-13   *Installing segmental-unit or stone first course.*

2. Adjust the granular base for the upper course so that it is level with the top of the lower block.

3. Set a second course of block straddling the base and the lower course. Center the blocks in the next course over the edge of the block below to create a staggered **surface pattern—a staggered, or running-bond pattern** (Figure 5-15). If necessary, move a block to one side to maintain the staggered pattern.

Adding additional courses:

1. Set a second course of block on top of the bottom course. If you are using a lean-back batter, the block should tilt at the same angle as the course below. If you are using a stepback batter, step the front of the second course back by approximately 1" from the lower course. Lipped-block will automatically create a setback batter when the lip slips behind the back of the lower course. If your wall alignment has curves, a portion of this lip may need to be removed to accommodate a tight turning radius. Carefully strike the lip with a claw

Figure 5-14 *Leveling and placement of the first course of segmental units.*

continued

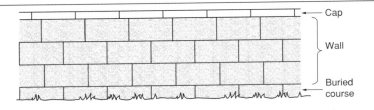

Cap

Wall

Buried course

Figure 5-15 *Staggered, or running–bond, surface pattern.*

hammer to remove a portion.

2. The joints between blocks on each course should be offset by half a block to create the staggered pattern shown in Figure 5-15. If necessary, slide a block to one side to maintain the proper alignment.

3. When the second course is completed, place a 4" diameter perforated drain tile behind the wall. Extend the tile along the entire length of the wall and run it around to the front of the wall at the low end (Figure 5-16), or through the front of the wall if that is the low point (Figure 5-4).

Figure 5-16 *Section of precast segmental-unit wall showing backfill and socked drain tile. This wall is being built with units that use fiberglass pins to help space and secure the wall.*

4. When you have completed the second course, set each subsequent course in a similar manner. Remember to stagger the joints and verify that the lips are locked to the lower course. Check to ensure that all courses are level and that there are no gaps between blocks.

5. After every course, fill any voids in the block centers with free-draining crushed rock. Sweep excess rock from the tops of the blocks.

6. After every course, backfill in the 12" zone immediately behind the wall with a layer of angular 1" crushed stone. The remainder of the backfill area can be soil excavated from the base.

Figure 5-17 *Cap installation for precast, segmental-unit walls. Some wall-material manufacturers make special caps to hide the wall structure below.*

Capping the wall:

1. When the setting and backfilling are complete, place the capstones on top of the wall following the staggered pattern used for laying the wall courses (Figure 5-17).

2. Some manufacturers recommend that a construction adhesive be used to secure the capstone to the top course.

Building a stone retaining wall

CAUTION

- Follow the manufacturer's instructions when using equipment.
- Use caution when cutting and installing wall materials.
- Obtain assistance when lifting wall materials.
- Do not construct any wall over 30" high without assistance from a qualified designer or contractor.

Time: Varies, depending on wall length and the amount of soil you have to excavate; approximately 1–2 hours for each linear foot of wall being built.

Level: Challenging (13 steps). Heavy lifting required.

Note: You should complete the base for your wall prior to installation. See Project: Installing The Base for a Retaining Wall presented earlier.

Tools Needed:

1. Round-nosed shovel.

2 Square-nosed shovel.

3. Wheelbarrow.

4. Utility knife.

5. Screwdriver.

6. Carpenter's level.

7. Rubber mallet.

8. Stone hammer.

9. Hand tamper.

10. Optional: **hydraulic stone cutter**.

11. Optional: vibratory plate compactor.

Materials Needed:

1. Enough stone to construct the wall. One ton of stone is typically enough to build 30 square feet of wall, but verify this number with your supplier. To calculate the amount of stone required, measure the square footage of the face of the wall (length times height, plus 1 foot). Divide the square footage by 30 to determine how many tons of stone are required.

2. Enough socked 4" perforated plastic drain tile to run the entire length of the wall, plus 5 feet. Purchase a T fitting if your drain tile runs through the face of your wall.

3. Free-draining angular 1" crushed stone. To calculate the quantity needed (in cubic yards), divide the square footage of the face of the wall by 27.

Directions:

Placement of the first course:

1. Beginning at the lowest end along the base, place a stone on the base material and aligned with front of the wall. Use the process described earlier to level a block (Figure 5-13, steps A through F). Using a carpenter's level, check to ensure that the stone is level end to end. If the stone is not level, add or remove base material from under one end of the material.

2. If you are employing a step-back batter, use a carpenter's level to check for level front to back. If the level is not correct, lean the stone backward and remove or add base material as necessary. If using a lean-back batter, place the stone with a 1/4" tilt from front to back.

3. Recheck for level in both directions after adjustments have been made.

4. Place the next stone with the end abutting the first stone. Place your hand on the joint between the stones to verify

continued

that the second stone is flush with the previous one. Adjust the second stone if it is too high or too low, and then proceed to level the stone the same way the first stone was leveled.

5. Continue to lay stone in this manner along the entire first course of the wall.

If the site requires that the wall steps up:

1. In places in which the wall needs to step up, complete the lower course of wall to the point at which the step occurs.

2. Adjust the granular base for the upper course so that it is level with the top of the lower stone.

3. Set a second course of stone straddling the base and the lower course. Do not align joints between courses where the wall steps up. If necessary, trim a stone or shift it to one side to maintain a staggered joint pattern.

Adding additional courses:

1. Set a second course of stone on top of the bottom course. If you are using a set-back batter, the front of the second course should step back approximately 1" from the front of the bottom course (Figure 5-18). If you are using a lean-back batter, the second course should lean at the same angle as the course below.

2. The joints between stones on each course should not be aligned with the joints of the course below. If necessary, select a different sized stone so that the joints can be staggered (Figure 5-19).

3. When the second course has been completed, place a 4" diameter perforated drain tile behind the wall. Extend the tile along the entire length of the wall and run it around to the front of the wall at the low end or through the front of the wall at the low point (Figure 5-4).

4. When the second course is complete, set each subsequent course in a similar manner. Remember to batter the stone and stagger the joints.

5. After every two courses, backfill in the 12" zone immediately behind the wall with a layer of angular 1" crushed stone. The remainder of the backfill area can be soil excavated from the base.

Figure 5-18 *Dry-laid stone wall showing set-back batter of 1" with each higher course.*

Figure 5-19 *The staggering of vertical joints in stone walls improves stability.*

Stair Installation

If you are planning a retaining wall of any substantial height, you will also probably need to plan some way to move efficiently from one level to another. To address this need, stairs can be included in your wall plan. Stairs have two primary components: the **riser**, or vertical portion of the stair, and the **tread**, or horizontal portion of the stair. For most stairs built as part of a wall, the thickness of the material will determine the riser height. Tread width can be increased by adding more material. For best installation, try to maintain a riser height of between 6" and 9", and a tread width of between 12" and 18".

Stairs can be constructed as part of a retaining-wall system by interlocking them into a **cheek wall**, or short-side wall—that is, the walls on each side of the stairs that will be at right angles to the main wall. This cheek wall helps hold the stair materials in place. Interlocking stairs should be constructed as the wall is being erected, with the first course of the stairs installed with the first course of the wall and continuing for each subsequent course. Construction techniques are similar for timber and stone installations, with segmental-unit stairs differing in the use of capstones set over wall material to create solid treads. Segmental units may pose difficulties in building stairs because of the necessity to trim blocks to fit your installation. If you need to cut and piece segmental units, consider having a contractor help you with the stair-installation process.

Building retaining-wall stairs

CAUTION

- Follow manufacturer's instructions when using equipment.
- Use caution when cutting and installing wall materials.
- Obtain assistance when lifting wall materials.
- Do not construct stairs with more than five risers without assistance from a qualified designer or contractor.

Time: Varies, depending on wall length and the amount of soil you need to excavate; approximately one hour for each tread being built.

Level: Challenging (12 steps). Heavy lifting required.

Note: Complete the base for your wall before beginning to install wall materials. See Project: Installing The Base for a Retaining Wall presented earlier.

Tools Needed:

Assemble the same tools necessary for constructing the wall in which the stairs are incorporated. Different lists will be required for timber, segmental-unit, and stone walls.

Materials Needed:

1. Enough wall materials (including spikes, adhesive, capstone, and other materials) to construct the stairs. Stairs will require three times the amount of wall material required for the same length section of wall. Perform the square footage calculations as for a wall, and triple the quantity for the short section where the stair is located

2. An additional 20 feet of socked 4" perforated plastic drain tile.

3. Free-draining angular 1" crushed stone, approximately 1 cubic yard (CY) for the stairs described.

Directions:

1. Using marking paint, mark the centerline location for the stairs.

2. Expand the granular base trench at the stair location to 24" further behind the wall base, and the width of the stairs plus 12" on each side. Fill this widened trench with base material and compact. When filled with base material the level should match that of the wall base.

continued

3. Place the first course of the wall across the opening for the stairs, keeping the front aligned.

4. At the location for the stairs, add additional rows of wall material behind the first wall course (Note: Timber stairs require two additional rows, but stone and segmental blocks may only need to add one additional row behind the first course.) The material must be the width of the stair opening plus 12" on either side to support the cheek walls (Figure 5-20, step A; for segmental-unit stairs, see Figure 5-21, step A.) The top of this additional material should be level with the wall course. This material will serve as the lower tread.

5. Route any tile around the back of this first course and lower tread. Continue the tile behind the wall after rerouting around the stairs.

6. Backfill and compact behind the wall. Behind the tread, backfill 12" further back. Also backfill the width of the stairs plus 12" on either side. If a tile is present, fill over the tile without disturbing its level. Smooth the base material so that it is flush with the top of the previous tread.

7. Construct the second course for the wall and stop short of the stair opening on each side of the stair.

8. Place material for the second tread on the base behind the first tread (Figure 5-20, step B; for segmental-unit stairs, see Figure 5.21, step B.). Shift the second tread forward to overlap the back of the first tread enough to create the desired tread width on the first step. This material must also extend 12" wider than the stair width on either side of the stair opening.

9. Create the cheek wall by placing wall material between the back of the wall and face of the second tread (for segmental walls the cheek wall will pass by the outside of the second tread). Trim the wall material as necessary to fit.

10. Backfill and compact behind the wall and tread.

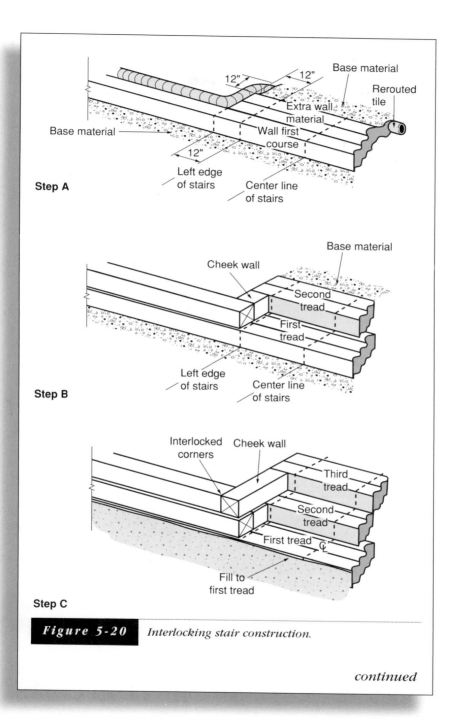

Step A

Base material

12"

12"

Base material

Extra wall material

Wall first course

Rerouted tile

Left edge of stairs

Center line of stairs

12"

Step B

Cheek wall

Base material

Second tread

First tread

Left edge of stairs

Center line of stairs

Step C

Interlocked corners

Cheek wall

Third tread

Second tread

First tread

℄

Fill to first tread

Figure 5-20 *Interlocking stair construction.*

continued

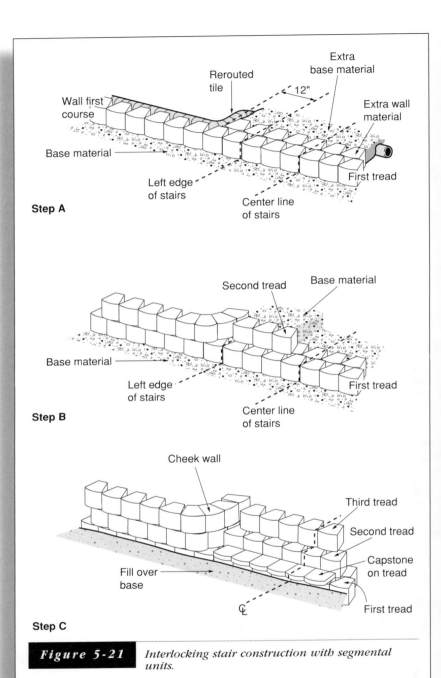

Step A

Step B

Step C

Figure 5-21 *Interlocking stair construction with segmental units.*

11. Repeat the above steps for each subsequent course of the wall and for each tread. Material at the intersection of the wall and the cheek wall should interlock, with each subsequent course overlapping the one below (Figure 5-20, step C segmental). For segmental unit stairs you can place a capstone on the tread to create a smooth surface. (see Figure 5-21, step C)

12. Fill in front of the first tread so that the grade matches the tread level.

Chapter 6

LANDSCAPE PAVING

Who doesn't like to soak up the sun on a beautiful patio or take a walk along the path to that secret niche in your garden? One of the most satisfying aspects of gardening is being able to use the spaces we have created, and one element that makes this possible is landscape paving. By providing surfacing we create drives, walkways, patios, entry areas, and a variety of functional spaces that can be enjoyed in all weather conditions. Consider using paving in your design in any of its valuable functions. Add a stone walkway, replace a deteriorating patio, redo the front entryway to match the exterior of your house. Paving in all forms can add function and beauty to a home.

Any paving material will work if it performs the essential function of separating your feet from the mud, but using paving in creative ways also enhances the constructed environment. In many projects the paved area is a wonderful source of esthetics, introducing color, texture, and materials that enhance the overall design of a project. Many paving choices available can be installed by a homeowner. Unit paving, such as **brick** and **concrete paving block**, dry-laid stone, and **granular paving** can be used for patios and walkways in many residential situations. These installations require time and effort but yield attractive and durable surfaces when successfully completed. Other paving types, including concrete and **masonry**, also provide exceptional surfaces, but these pavement methods typically require the special skills and expertise of the landscape contractor. Your choice of paving materials will require balancing design esthetics and your installation skills.

PLANNING AND PREPARATION FOR LANDSCAPE PAVING

Preparation for a paving project requires consideration of several site-related situations and an understanding of the installation methods for the selected material. Base preparation will be similar for all pavement methods, but the specifics of installation, such as patterns, edging, and finishing, will vary from one pavement to another. Carefully review the planning consideration noted below to determine whether it applies to your project. Some municipalities may restrict the amount of impermeable paving that can be installed, so that water runoff is reduced. Check with local officials before you begin, in order to avoid a mid-project change in plans.

Access to the Site and Delivery of Materials

Unless you plan to carry bricks by hand, you will need access to the site for motorized vehicles delivering base, paving, and edging materials. Plan a route that will provide access to the site for large equipment such as a delivery truck and for small equipment such as a skidsteer. Delivery of unit pavers close to the site typically requires access by a large truck with a **loading boom**. Delivery of granular paving will require access to the site by a dump truck or by a skid steer loader moving material from a stockpile. If possible, this access route should not cross over the prepared paving area. Plan the installation so that the pavers and other materials can be delivered and the excess removed without damaging a completed installation. This will save you time and money.

After your base has been completed and you have begun a paving installation, the highest quality and efficiency is achieved by moving through the project without pause. This requires that materials be ready for installation and that any equipment necessary for installation or cutting be on hand at the beginning of the project. A staging area where several pieces of material can be laid out will speed the selection process required when laying stone. It will also be beneficial to locate an area where additional paving and granular material can be stockpiled, so that you can refurbish the surface in the future, if necessary.

Project Layout

Measure and mark project locations to guide your initial excavation. Painting the limits of your project is easier than dodging strings and flags, because both people and vehicles can pass directly over a painted line without damaging it or becoming entangled in it. Mark and excavate an area that goes at least 2' beyond the edge of the proposed paved area to create an **apron** that can be used for construction and later used to gradually match the existing grade to the new paving. This apron also provides the space to extend base material 1' beyond the edge of the project to support edge restraints. If the project edge is adjacent to a steep slope, a wider apron may be required in order to blend grades. This situation may also require installing a wall to support the paving; if so, install the wall before proceeding with any paving.

Excavation and Subgrade Compaction

Because paving creates relatively flat surfaces with slight slopes introduced for drainage, proper grade is important. Attempting to work with slopes that are too steep will make installation difficult and may result in surfaces that are too slippery to be safe. Slopes that are too flat will result in the ponding of water that does not adequately drain from the surface. A **cross-slope** of 2 percent (1/4" fall for every 1' of horizontal distance) is recommended for most paving. Cross-slopes (an entire surface that slopes in only one direction) are the easiest to construct and the best for drainage. The slope on paving surfaces should always be directed away from structures and permanent improvements.

Using the most efficient means available, dig out excess soil until it is excavated to a uniform depth over the entire area to be paved. The base excavation should extend to the markings. To determine how deep you should excavate, add the paver thickness (anywhere from 1 1/2" to 3"), the depth of any sand setting bed (typically 1"), and the thickness of the base (typically 4"). For many paving surfaces, this will be approximately 8", but each dimension should be checked with materials suppliers and your design before excavation begins.

To check the excavation for proper depth while you are digging, place a **string line** from one side of the project to the other that is set at the same height as the proposed pavement surface, including the cross-

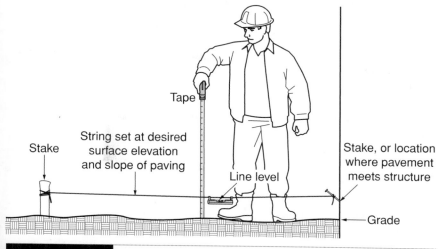

Figure 6-1   *Checking the cross-slope and depth of pavement excavation.*

slope. As you are excavating, measure the excavation depth from the string (Figure 6-1). Any vegetation and unsuitable soils must be removed. Areas that are over excavated should be filled with angular 1" crushed stone and compacted. When the excavation is complete, verify that the base is the proper depth and that it slopes in the direction you desire.

Undisturbed earth is preferable for **subgrade**, or the soil below the base. Because it is already compacted you can accurately determine the stabilized height of materials and construction that will rest on it. If a paving installation is constructed in an area in which the subgrade has been disturbed, proper compaction will be necessary. Compacting subgrade in paved areas of up to 1000 SF can be done with a hand tamper or using a vibratory plate compactor (vibraplate). Utility trenches under paved areas should be compacted in 6" lifts as the trench is backfilled. Small areas that cannot be accessed by a vibraplate should be hand tamped. An informal test of the base for firmness is to observe whether foot traffic or a loaded wheelbarrow passing over the surface leaves any rutting or depressions. Such conditions indicate incomplete or improper compaction or an unstable subgrade. Any poor soils encountered must be corrected by removal and replacement with suitable material. See the project: Fixing poor soil problems below your project, Chapter 4.

Utility and Related Preliminary Work

Because paving is not easily changed, it is important to ensure that other work that might impact an installation be completed before paving begins; for example, if walls are necessary to create the area for paving, they should be completed and thoroughly compacted before you begin paving. All underground utility lines in the area should be in place and trenches below the area paved, backfilled, and compacted. Empty conduit or piping can be placed under the paved area if future utility installations, such as lighting or irrigation, are anticipated.

Base-Material Installation

Durable paving installations can typically be traced to a sound and stable base. Although there is a temptation to place pavement directly on soil, the quality and life of the project will be greatly enhanced by taking the time to prepare the area below the pavement with the same care as that applied to paving the surface. Adequate base preparation requires excavation of the site, examination of the subgrade for stability, correction of any problems, and finally placement and compaction of a suitable base material. Performing these steps carefully lays the foundation for a long-lasting pavement installation.

Installing the base for a paved surface

CAUTION

■ Verify the location of all utility lines prior to construction.

■ Follow the manufacturer's instructions when using equipment.

Time: 2–4 hours for every 100 SF of base prepared.

Level: Moderate (8 steps). Digging required.

Tools Needed:

1. Plan of project.

2. Marking paint.

3. Mason's twine.

4. Line level.

5. Four 2" × 2" × 2' long wood stakes.

6. 2-pound sledge.

7. 25' measuring tape.

8. Round-nosed shovel.

9. Square-nosed shovel.

10. Garden rake.

11. Wheelbarrow.

12. 10' long 2 × 4.

13. Hand tamper.

14. 10' long 1" diameter metal pipe.

15. Location for disposing of excavated soil.

16. Optional: Vibratory plate compactor.

Materials Needed:

1. Angular 1" crushed stone, approximately 1 CF for every 2 SF of base area for a 4" thick base.

Directions:

1. Use the paint to mark the site to be paved. The excavation for the base should extend a minimum of 2' outside the edge of the proposed paved area, but the base should be installed only 1' beyond the edge of the proposed project. The placement of base material may be made easier if the project edge is repainted after the excavation is completed.

continued

2. Remove and dispose of all vegetative cover. Excavate and remove the remaining roots of larger plants.

3. Verify the excavation depth by adding the paver thickness, setting bed thickness (1"), and base thickness (4"). Establish a stringline (Figure 6-1), set at the desired finish height of the paving, including any cross-slope. If significant excavation is required, dig to the approximate depth of the base before installing stringline.

4. Excavate to the calculated depth over the entire base area (Figure 6-2). Check the depth of your excavation often by measuring down from the stringline.

5. Examine the subgrade for potential soil and water problems. If corrections are required, refer to Project: Fixing poor soil problems below your project, Chapter 4.

6. Place a 4 1/2" thick layer of angular 1" crushed stone over the excavated area (Figure 6-3); the layer will compact to approximately 4" thick.

7. Compact the base using a hand tamper or a vibratory plate compactor. Work from the outside edges into the center, and then repeat at an angle to the direction of the first pass.

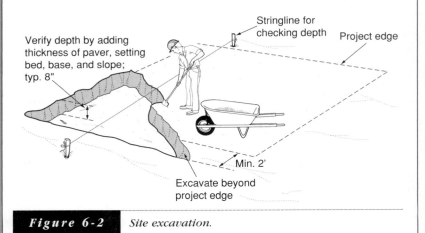

Stringline for checking depth

Project edge

Verify depth by adding thickness of paver, setting bed, base, and slope; typ. 8"

Min. 2'

Excavate beyond project edge

Figure 6-2   *Site excavation.*

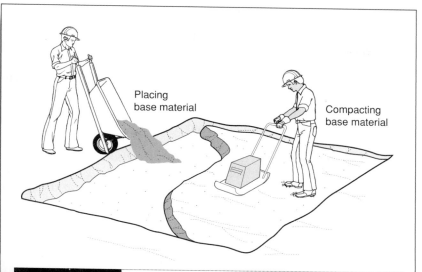

Placing
base material

Compacting
base material

Figure 6-3   *Placement and compaction of granular base.*

8. Drag a 2 × 4 across the base and level the material to within 3/8" of the desired grade, using the stringline as your guide. Check for irregularities using a long pipe rolled over the surface. Paint around low and high areas. Add base material to low areas and recompact them. Use a rake to remove high areas (Figure 6-4).

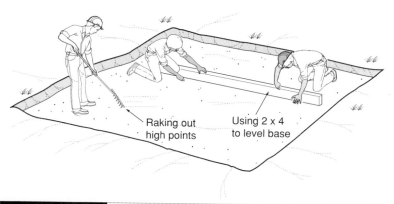

Raking out
high points

Using 2 x 4
to level base

Figure 6-4   *Leveling the base.*

Pavement Patterns

The beauty of many pavement types (brick, concrete paving block, stone) is enriched by the pattern used to arrange the individual pavers. A feel of craftsmanship is created by employing patterns that require more expertise to install than simply leveling a granular material. Many pattern choices are available, but you will find that most are variations or combinations of a few simple arrangements. Select a pattern and practice the placement of units within that pattern to determine whether that look is right for your project. Note that some patterns, such as **herringbone** and **running bond,** require more cutting than **stacked bond** and **basket weave.** Herringbone will also require constant monitoring of individual paver placement to maintain the pattern.

Stone paving requires a different approach to installation than unit pavers. The installation sequence for unit pavers is repeated over and over, because of their uniform shape. Stone paving units, however, can be irregular and inconsistent in size and shape, requiring you to examine each piece for fit with the pattern. You must not only view the piece from an esthetic perspective but also locate or trim pieces to create a stable installation.

Paving patterns

CAUTION

■ Use caution when splitting paving material. Wear appropriate safety equipment, including gloves and safety glasses.

Time: 1–2 hours.

Level: Easy (10 steps).

Tools Needed:

1. Chalk line.

2. A flat 5' × 5' section paved area for practice. No base is required.

3. Brick hammer.

4. Brick set.

Materials Needed:

1. Paving material for practice. Twenty full pavers and 5 half pavers for each different material you plan to practice with. If practicing stone paving provide 10 pieces of stone paving.

Directions:

Practice for unit paver patterns:

1. Locate a beginning point for the pattern. This point should be along a straight edge.

2. Snap a chalk line that is perpendicular to the straight edge. Most patterns will begin along the straight edge and move outward along this chalk line, filling in diagonally.

3. Begin placing blocks according to the numbered sequence shown for each pattern:

   - Stacked bond (Figure 6-5).
   - Running bond (Figure 6-6).
   - Basket weave (Figure 6-7).
   - Herringbone (Figure 6-8).

Practice for stone paving pattern:

1. Locate a beginning point for the pattern. This point should be located along a straight edge or in a square corner.

2. Begin placement of stone in the direction shown. This sequence will be based on the size and shape of stone available and is intended only to illustrate the selection and placement of stone. Work across the surface one row at a time.

   - **Random irregular** (Figure 6-9).

continued

A. Consider installing soldier course around edge
B. After placing paver 17, continue stacking diagonal pattern until surface is covered
C. Cut pavers if necessary to fit edges of paved area

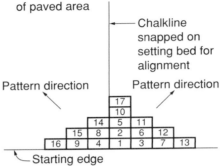

Chalkline snapped on setting bed for alignment

Pattern direction   Pattern direction

Starting edge

Figure 6-5 *Stacked-bond-pattern installation.*

A. Consider installing soldier course around edge
B. After placing paver 13, continue pattern until surface is covered
C. Cut pavers if necessary to fit edges of paved area
Note: Pavers in alternate courses (pavers 4,9,12,...) are half pavers

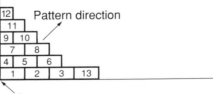

Pattern direction

Square corner

Starting edges must be at right angles

Figure 6-6 *Running-bond-pattern installation.*

A. Consider installing soldier course
   around edge
B. After placing paver 18, continue
   pattern until surface is covered
C. Cut pavers if necessary to fit edges
   of paved area

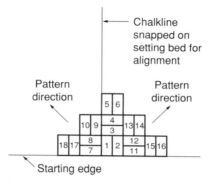

Chalkline snapped on setting bed for alignment

Pattern direction

Pattern direction

Starting edge

Figure 6-7 *Basket-weave-pattern installation.*

A. Consider installing soldier course
   around edge
B. After placing paver 30, continue
   pattern until surface is covered
C. Cut pavers if necessary to fit edges
   of paved area
Note: Every third paver along edge
   (pavers 3, 4, 9, 17, 28, 29,…) is
   a half paver

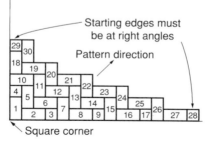

Starting edges must
be at right angles

Pattern direction

Square corner

Figure 6-8 *Herringbone-pattern installation.*

continued

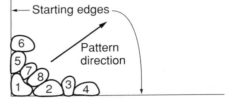

A. Begin with stone that matches corner angle
B. Lay straight-edged stones along sides
C. Fill center with stones that create three friction points with adjacent stone and joints 1/2" or less
D. Adjust stones so they are flush with adjacent stones
E. Mix small and large stones randomly in pattern; avoid aligning stones with straight joints

Starting edges

Pattern direction

6 5 7 8 1 2 3 4

Figure 6-9 *Random-irregular-pattern installation.*

3. Use stone with straight sides along pavement edges.

4. Stone should make contact with adjacent stones in three places and should leave joints between stones of _" or less.

5. Replace stone or adjust placement to achieve the desired pattern.

Splitting and shaping paving material

Certain paving patterns will require the installer to use portions of the paver to complete the pattern. For some paving materials, halfs and starter pieces can be purchased. In other situations, the paver must be split, or **cleaved**, by hand. Clay brick can be split using a **brick set**, a chisel-like tool with a wide blade, designed to separate the material along natural cleaving planes. Concrete paving block can also be cut using this method but does not always cleave evenly. Stone can be cut in a similar manner as clay brick, although there may be significant irregularities in the edges of the stone after being cleaved. As with any paving product, it will take practice to obtain good splits. Purchase extra material, so that mistakes will not leave you short of paving material for your project. If the material you are paving with is impossible to split, consider having a contractor cut the shapes you need using a **wet masonry saw**. If sharp edges of an individual stone or paver need to be removed to improve fit, shaping can be done using a **brick hammer**.

Splitting paving materials using a brick set

CAUTION

- When splitting any material, wear proper clothing and safety equipment, including gloves and safety glasses.
- Use all equipment according to the manufacturer's instructions.

Time: 15 minutes

Level: Easy (7 steps).

Tools Needed:

1. Brick hammer.

2. Brick set.

Materials Needed:

1. Materials to split. Gather a collection of unit pavers and stone types to practice with.

2. Bed of sand on which to set pavers.

continued

Directions for splitting paving material:

1. Using the brick set, lightly strike the paver on all sides at the location where the material is to be cut.

2. Set the paver flat on the sand bed. Any solid surface will also work.

3. Align the brick set along the mark for the cut.

4. Strike the end of the set with a hammer. The brick should cleave at that point (Figure 6-10).

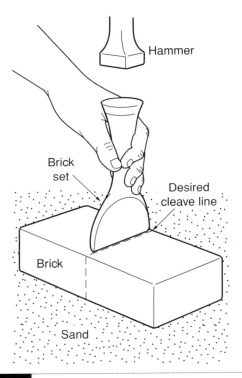

| **Figure 6-10** | *Cleaving a brick with a set.* |

Directions for shaping stone:

1. Hold the stone firmly.

2. To reduce irregular splitting of the stone, strike the protruding piece from the side rather than the top (Figure 6-11).

3. Strike the stone with the blunt end of the brick hammer.

Figure 6-11 *Shaping a stone edge with a brick hammer.*

Setting Bed For Unit Pavers

Pavements such as brick, concrete paving block, and dry-laid stone require a special layer of material between the base and paver to ensure that the surface is smooth and level. This layer, called the **setting bed**, is typically a 1" layer of sand that is smoothed by a process called **screeding**. To screed the setting bed, place two 1" diameter pipes on your leveled base (Figure 6-12). Mound sand on the base and spread it around the pipes. Place a straight 2 × 4 across both pipes. Using a sawing motion, pull the 2 × 4 across the surface, smoothing the sand as you go. The 2 × 4 should always maintain contact with the pipes. If voids occur behind the screed, fill with sand and rescreed. When finished, carefully lift the pipes out of the setting bed and fill the voids with sand. You can smooth any irregularities using a **magnesium float**, a flat-bottomed tool used for concrete finishing.

The setting bed removes any irregularities introduced during base installation and provides a surface on which the paving material can be

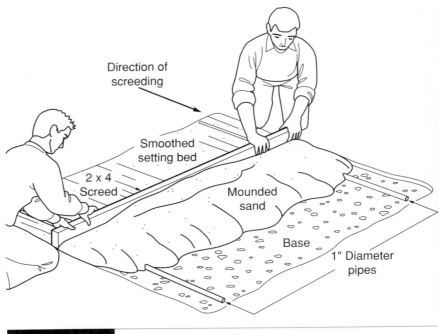

Direction of
screeding

Smoothed
setting bed

2 x 4
Screed

Mounded
sand

Base

1" Diameter
pipes

Figure 6-12   *Screeding a setting bed for paving. The 2 × 4 slides across
the 1" pipe, leaving a smooth surface.*

easily set. The setting bed is typically placed and screeded just before
the beginning of paving placement. If paving a large area, screed small
areas of setting bed, lay pavement in that area, and then proceed to the
next location. The sand in the setting bed is easier to work if it is lightly
misted before placing pavers. Make the misting very light, so that the
finished surface is not disturbed.

Edge Restraints For Paved Surfaces

Most paving materials require that the edges be held in place to prevent shifting and settling. The material that holds this paving in place
is called **edge restraint**. Of the many methods that effectively restrain
paving, preformed plastic edge restraints are the most common. You
can place edge restraints in position, push them firmly against the

paving, and anchor them using stakes or nails. You can place plastic edge restraints on top of the setting bed or remove the setting bed and anchor the restraint to the base. If your paving material is not held in place by a structure or other paved surface, you will need to install the edge restraint to ensure that your project will not deteriorate over time. (Figure 9-19)

In addition to preformed plastic you can use wood, concrete curbs, metal, or other materials to hold paving in place. These other types of edge restraint may require excavation and placement before you pave, or they can be installed after paving in a manner similar to that used with preformed plastic.

Installing Landscape Paving

Actual installation of the paving material is the last in this series of steps. Proper preparation of the site and the construction of a sound base will make installing the pavement surface easier. If all preparation has been done properly, completion of the surface should proceed smoothly and quickly.

Unit Pavers

Unit paver is a collective term that includes clay bricks, interlocking concrete pavers, adobe pavers, precast concrete units, and similar types of materials that are produced with consistent dimensions and installed as individual pieces. Unit pavers are among the most desirable and requested types of paving materials available because of the esthetics of the finished surface. This attractiveness is not without its price, however, because unit pavers are high in cost and require skill and effort to install.

Although the strength of properly installed unit-paver surfaces is very good, failure of base or subgrade under pavers can create safety and long-term maintenance problems. All unit paver types have numerous joints between the pavers, and even after a perfect installation, these joints could hinder snow removal. Poorly installed unit pavers can create tripping hazards if they shift. Most residential installations are intended for patios, walks, and areas that support predominantly foot traffic, but with an engineered base course, concrete paving block and brick can be used in areas in which vehicular traffic is expected.

Installing a patio using bricks or concrete pavers

CAUTION

■ Use caution when cutting materials.

■ Follow the manufacturer's instructions when using equipment.

Time: Varies based on size of project, 6–8 hours for a 10' square patio with minimal cutting.

Level: Moderate (23 steps). Equipment operation, heavy lifting, and excavating required.

Note: To install a durable pavement, a base course should be installed under your pavers. To install a base course, review the project titled "Installing the base for a paved surface". If pavers need to be cut, review the project titled "Splitting paving materials", or hire a contractor who has a wet-masonry saw.

Tools Needed:

1. Plan of project.

2. Marking paint.

3. Roll of mason's twine.

4. 25' tape measure.

5. Round-nosed shovel.

6. Square-nosed shovel.

7. Garden rake.

8. Broom.

9. Wheelbarrow.

10. 10' long 2 × 4.

11. Two-pound sledge.

12. Vibratory plate compactor with rubber boot covering.

13. Two 10' long 1" diameter pipes made of metal or PVC.

14. Magnesium concrete float.

15. Rubber mallet.

16. Tin snips or hacksaw.

17. Torpedo level.

18. Screwdriver.

19. Carpenter's square.

Materials Needed:

1. Brick or concrete paving block approximately 4.75 bricks per square foot of patio surface. Concrete paving block quantities will vary in size, so confirm with your materials supplier. Order 5% extra for cutting and breakage.

2. Clean, coarse, and damp concrete sand or stone dust, enough sand or stone dust to place a 1" layer over the entire base area with additional concrete sand to sweep into joints after paving; approximately 2.0 CF for every 10 SF of paved area. Use volume formulas from Chapter 2 to calculate the amount of sand required.

3. Edge restraint, 1 linear foot for each foot of perimeter of patio or walkway you plan.

4. 12" × 3/8" spikes or edging stakes, one for each linear foot of perimeter of patio or walkway.

Directions:

Placement of setting bed:

1. Set two 1" diameter pipes directly on base.

continued

2. Spread the sand evenly over the pipes and base. Spray dry sand lightly with water before screeding.

3. Set the 2 × 4 on top of the pipes and level the sand by dragging the 2 × 4 along the pipes (this process is called screeding). Using a sawing motion while dragging the 2 × 4 will help speed the process. Screed perpendicular to the rails when possible.

4. Screed only the area of setting bed that will be paved immediately. You may need to screed small sections to match existing grades.

5. When screeding is completed, lift out any temporary screed rails. Fill the voids left by the rails with additional sand, and smooth the filled areas to match the surrounding base with a magnesium float (Figure 6-12).

Setting pavers:

1. Select and practice a paving pattern.

2. Install a **soldier course** around the edge of the area to be paved. If the entire area is not yet screeded, place the soldier course in the area that is screeded (Figure 6-13).

Figure 6-13   *Soldier course at the edge of pavement. This course stabilizes the pavement by reducing the number of small pieces at the edge.*

3. Identify a location to begin the paving. The center of a straight edge of the project is the best beginning point. Snap a chalk line perpendicular to the straight line on the setting bed to guide the placement of the first pavers (use the carpenter's square as a guide). A square corner can also be used to begin pavement patterns, but the corner must be truly square to avoid "pinching" the pattern.

4. Begin placing pavers according to the steps for the selected pattern (Figures 6-5 through 6-8).

5. Place the pavers using a straight-down motion, and do not twist or turn them once set on the setting bed (Figure 6-14). Place pavers with their sides in contact with each other. If a paver is too high or low, carefully lift it and add or remove sand below it. As the pattern moves outward, kneel on the set pavers to reach the edge. Avoid stepping or leaning on the pavement within one foot of an unrestrained edge.

Figure 6-14

Beginning the placement of interlocking pavers. The stringline runs along the center of the paving pattern. A chalkline can also be snapped on the setting bed rather than using a stringline.

6. Continue placing pavers, working across the interior of the area to be paved.

7. As the pavers fill the area, continue the soldier course of full pavers along the entire perimeter before beginning to install the pattern on the interior.

continued

8. Check alignment of pavers using a string-line or straightedge. Adjust the position of the blocks using a putty knife or prybar.

9. Continue placing pavers in the desired pattern. Split partial pavers when required. Cut pavers may have to be gently tapped into place using a rubber mallet. An alternative to cutting half or partial pavers while the pattern is being installed is to place all full pavers first, then mark, split, and place partial pavers as the last step in the process. This will save time, because it will enable you to do all of the cutting required at once. (Figure 6-15).

Figure 6-15

Setting cut pavers. Tapping with a rubber mallet may be necessary to push pavers into a small opening. Pavers should fit snugly, with no wide joints.

Edging and seating pavers:

1. When all pavers have been placed, measure and cut a piece of edge restraint for each side of the paved area.

2. Carefully remove the sand setting bed directly under the edge of the paver. Place the edge restraint directly on the base (Figure 6-16). The edge restraint should extend a minimum of half-way up the thickness of the paver (Figure 6-19).

3. Hold the edge restraint in place, and, using the two-pound sledge, drive 12" × 3/8" spikes or edging stakes through the edge restraint and into the base.

4. Place the spikes at the beginning and end of every edge restraint piece, and every 12" around the edge of the paved area; this will prevent paving blocks from drifting outward from the paved area because of the action of freezing and thawing, running water, foot traffic, or other forces that may act upon them.

Figure 6-16

Installation of plastic edging after placement of pavers. Tap the edge restraint lightly with the hammer to avoid lifting the pavers.

5. Backfill around the edge of the paved area. If planting the area around the paving use a material suitable for growing.

6. Sweep dry concrete sand into the joints of the pavers.

7. Sweep any stone or debris from the paved area.

8. Set a vibratory plate compactor with a rubber boot on the paved area, start it, and run it around the edge and over the entire surface.

9. Again sweep dry sand into the joints of the pavement to fill any voids that may have been created by sand settling into or flying out of the joints between the blocks because of the vibration of the compactor.

Stone

The intricate textures and patterns of limestone, bluestone, granite, or slate patios result in a highly attractive paving surface. Balanced with the esthetics of stone are the higher costs and skill levels required for its installation. The availability of materials will have an impact on price, resulting in lower costs in areas in which stone is commercially quarried and available, and higher prices further away. Stone also carries with it concerns about stability and safety. Stone paving has surfaces and joints that can be irregular, creating the potential for tripping and making snow removal difficult. Any base failure or water below the paving could force the stone out of position, amplifying the problems. When properly installed, however, few other surfacing materials produce as attractive a finished private walk or patio as stone.

Installing a stone patio

CAUTION

- Use caution when splitting materials.
- Follow the manufacturer's instructions when using equipment.

Time: Varies, based on size of project; 8–10 hours for a 10' square patio with minimal cutting.

Level: Moderate (20 steps). Heavy lifting and excavating required.

Note: For a durable pavement, install a base course under your patio. To install a base course, review the project titled "Installing the base for a paved surface" presented earlier. If stone needs to be cut, review the project titled "Splitting paving materials", or hire a contractor who has a wet masonry saw.

Tools Needed:

1. Plan of your project.

2. Marking paint.

3. 25' tape measure.

4. Round-nosed shovel.

5. Square-nosed shovel.

6. Garden rake.

7. Broom.

8. Wheelbarrow.

9. 10' long 2 × 4.

10. Rubber mallet.

11. Stone hammer.

12. Two 10' long 1" diameter pipes made of metal or PVC.

13. Magnesium float.

14. Torpedo level.

Materials Needed:

1. Stone paving, approximately 1 ton of stone for every 30 square feet (SF) of paved area. Check with stone supplier for quantities required. Order 5% extra for cutting and breakage

2. Clean, coarse, and damp concrete sand or stone dust, Enough to place a 1" layer over the entire base area and to sweep between joints after paving; approximately 2.0 cubic foot (CF) for every 10 SF of paved area. Use volume formulas from Chapter 1 to calculate the amount of sand required

3. One bag mortar mix for every 10 SF of paved area.

4. Edge restrain. 1 linear foot (LF) for each foot of patio perimeter.

5. 12" × 3/8" spikes or edging stakes. One for each LF of patio perimeter.

continued

Directions:

Placement of setting bed:

1. Set two 1" diameter pipes directly on base.

2. Spread the sand evenly over the pipes and base. Dry sand will need to be sprayed lightly with water before screeding.

3. Set the 2 × 4 on top of the pipes and level the sand by dragging the 2 × 4 along them. A sawing motion while dragging the 2 × 4 will help speed the process. Screed perpendicular to the rails whenever possible.

4. Screed only the area of the setting bed that will be immediately paved. Small sections may need to be screeded to match existing grades.

5. When screeding is completed, lift out any temporary screed rails, fill the voids left by the rails with additional sand, and smooth the fills to match the surrounding base, using a magnesium float.

Setting stone:

1. Lay out stone before placing it on the setting bed to practice the pattern. Use large stones for all edges, and alternate large and small stones throughout the remainder of the project.

2. Identify a location in which to begin the paving. A corner of the project is a good beginning point.

3. Set the stone directly down, and do not twist or turn them once set on the setting bed. Place stones so that joints are less than 1/2" between stones and that each stone makes contact at three points with each neighboring stone. If necessary, shape stones with a stone hammer. If a stone sets too high or low, carefully lift the stone and add or remove sand below (Figure 6-17).

Figure 6-17 *Proper placement of dry-laid stone.*

4. Continue placing stones, working along a diagonal moving out from the beginning point.

5. Check to ensure that each stone is flush with surrounding stones when in place. If it is not flush, lift it and add or remove sand as necessary.

6. Stones with irregular edges can be shaped slightly by striking with a stone hammer.

7. Avoid stepping within a foot of an unrestrained edge.

8. In areas in which a partial stone is required, mark a stone and split it (Figure 6-18). Install the split stone in the opening. Tap lightly using a rubber mallet if necessary to fit the stone fragment into position.

continued

Edging stone:

1. When all stone is placed, measure and cut a piece of edge restraint for each side of the paved area.

2. Carefully remove the sand setting bed directly under the edge of the stone. Place the edge restraint directly on the base The edge restraint should extend high enough to cover a minimum of half the thickness of the stone (Figure 6-19).

Figure 6-18

Splitting stone with a brick set. Set the paving material on sand or a flat surface. Place the blade of the set along the line on which you want the stone to cleave, and then strike firmly with a hammer. Trimming small amounts at a time can be more effective than cleaving large portions.

3. Holding the edge restraint in place, drive 12" × 3/8" spikes or edging stakes through the edge restraint and into the base.

4. Place the spikes or stakes at the beginning and end of every edge restraint piece, flush against the edge, and every 12" around the edge of the paved area.

5. Backfill around the edge of the paved area. If planting next to the paving, fill with a material suitable for plant growth.

6. Sweep a mixture of 50 percent dry concrete sand and 50 percent mortar mix into the joints of the stone.

7. Mist the paving with water.

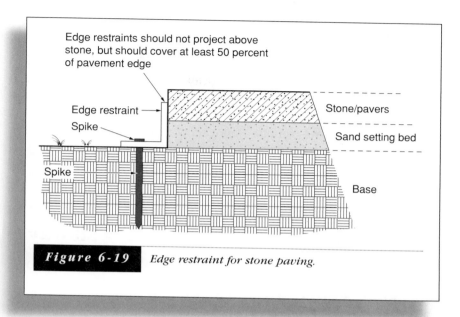

Edge restraints should not project above stone, but should cover at least 50 percent of pavement edge

Edge restraint

Spike

Spike

Stone/pavers

Sand setting bed

Base

Figure 6-19 *Edge restraint for stone paving.*

Granular Paving

Granular surfaces are typically found in drives, walkways, trails, and occasionally in outdoor living areas. Granular paving includes such materials as crushed stone, crushed brick, pea gravel, decomposed granite, and other permanent materials that are available as small pieces. You will want to consider carefully the location where you plan to place your pavement and the use to which it will be put before deciding on the material to be used; for example, the limited durability of thin applications of granular paving will limit them to locations that are primarily foot traffic with limited vehicular traffic. Granular pavements can be quite attractive, because they provide consistent texture and color, but installing them does not require a high level of craftsmanship.

Installation is easy and costs typically low for granular materials, advantages that recommend their use in residential settings. Granulars with permanent edgings such as stone, brick, or concrete can create useful walkways and outdoor living areas. However, some materials tend to be subject to such safety hazards as washouts, poor traction, and sharp particle edges. Granular surfaces tend to be quite labor intensive after the initial installation. Maintenance includes the constant need to maintain the edging, frequently level, clean up material that sticks to shoes and is tracked inside, and add more material to replace that removed by foot traffic, wind and rain erosion. Granulars similarly are not a recommended choice in areas in which snow removal is necessary, because snowblowers or shovels tend to remove a portion of the material each time they pass over the surface.

Installing a granular-material walkway

Time: 1–2 hours for every 10' of walkway.

Level: Moderate (7 steps). Lifting and excavating required.

Tools Needed:

1. Plan of your project.

2. Marking paint.

3. 25' measuring tape.

4. Torpedo level.

5. Round-nosed shovel.

6. Square-nosed shovel.

7. Broom.

8. Wheelbarrow.

9. Garden rake.

10. Sod roller.

11. Optional: other edge restraint methods may require additional tools to cut and install the material.

Materials Needed:

1. Finely crushed granular material (that which passes through a 3/8" sieve). Every 10" of a 5' wide walkway will require approximately 18 CF of material.

2. Edge restraint. 2 LF for every 1 LF of walkway installed; if edging a patio, 1 LF for every LF of perimeter.

3. 12" × 3/8" spikes or edging stakes, one for each linear foot of perimeter of patio or walkway.

4. Disposal area for excavated soil.

Directions:

1. Review the alignment of the pathway.

2. Mark both edges of the pathway using the paint.

3. Using a shovel, excavate the pathway area to a depth of 4". If areas of poor soil are encountered, correct the problem using the instructions found in the project "Fixing poor soil problems under your project".

4. Install edge restraint along both sides of walkway alignment. If cross-slope is important, place 2 × 4 across walkway and

Figure 6-20

Placement of granular paving material.

continued

slope using torpedo level.

5. Backfill along outside edges of edge restraint. If planting along walkway fill with a material suitable for plant growth.

6. Place a 2" layer of granular material in the excavated area of the pathway (Figure 6-20). Use the sod roller to compact this layer (Figure 6-21).

7. Apply another 2" layer of granular material. Rake level and compact again. Material should be level with the top of the edge restraint.

Figure 6-21

Leveling and finishing granular-paving installation.

Stepping Stones

Informal walkways can be constructed with loose pieces of paving arranged in a convenient stepping pattern. **Stepping stones** can be composed of any paving material, with **flagstone** used most often for informal walkway applications. These unmortared stepping stones placed on earthen or thin granular bases are inexpensive but are suitable primarily for private walks; areas with heavy traffic are not good candidates for this type of surface.

Installing stepping stones

Time: 15 minutes for each stepping stone.

Level: Moderate (7 steps). Light lifting required.

Tools Needed:

1. Plan for the project.

2. Flour.

3. Round-nosed shovel.

4. Square-nosed shovel.

5. Wheelbarrow.

6. Hand trowel.

7. Broom.

Materials Needed:

1. Stepping stones, each approximately 12" × 12" × 2" thick; one stone for each 18" on path desired.

2. 2 CF of sand for each 20 LF of walkway.

3. Disposal location for excavated materials.

Directions:

1. Review the plan for the pathway.

2. Lay out the stones in a comfortable stepping pattern (Figure 6-22, Step A).

3. When layout is acceptable, sprinkle flour around the edge of each stone to mark its outline (Figure 6-22, Step B).

continued

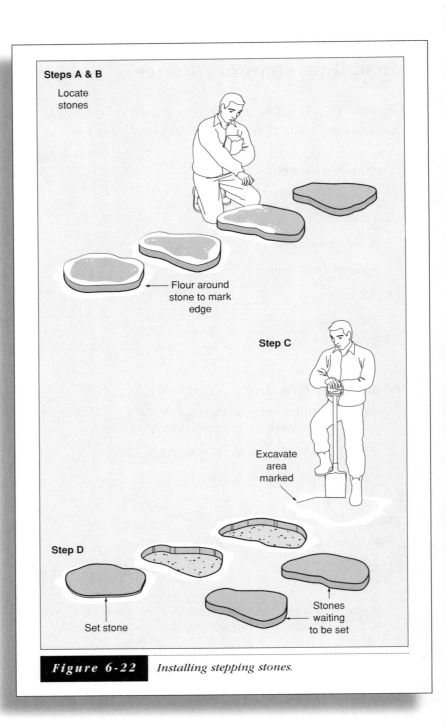

Steps A & B

Locate stones

Flour around stone to mark edge

Step C

Excavate area marked

Step D

Set stone

Stones waiting to be set

Figure 6-22 *Installing stepping stones.*

4. Remove each stone and excavate a 1" deep base inside the outline for each stone (Figure 6-22, Step C). Properly dispose of any plant material or soil excavated. Set the stone back in the base and adjust with the hand trowel if necessary (Figure 6-22, Step D).

5. If a stone does not sit level and stable, adjust using a small amount of sand under the stone.

6. Repeat the setting process for each stone.

7. Sweep stone surfaces clean.

Chapter 7

WOODEN STRUCTURES

Despite spending most of their lives indoors, when people find the time to relax outdoors they still gravitate toward structures. From inside the gazebo, we bask in the glories of nature. We relax in the hammock strung between trees or the posts of a shade structure. From a vantage point on our **deck,** we watch the children play. Regardless of the beauty of garden, the move toward structures is inexorable. Structures shelter us from the sun, protect us from attacking insects, warm us from the chill, and serve the general functions that we enjoy indoors, but with the feel that we are closer to nature by virtue of not being completely enclosed. Wood structures provide practical, attractive, and usable space in the landscape.

Whether your structure is a multilevel deck or a simple **trellis,** the presence of a structure adds warmth, interest, and, in many designs, fulfills the important design principle of scale. Unlike paving and ground-level hardscape improvements, structures instantly change the dynamics of the design by introducing three-dimensional elements. Regardless of the function, adding a crafted wooden element brings another dimension to the landscape.

Several types of wooden structures have proven to be valuable additions to the landscape. Decks are surfaced platforms that serve several functions, including entry, grade transition, and entertainment. Enclosed structures such as **gazebos** serve as climate-controlled outdoor rooms. **Arbors**, or roofed structures with open sidewalls, provide both enclosure and openness (Figure 7-1). Simpler structures, such as trellises, can be built to support vines or to create shade for outdoor living areas.

The construction of wooden structures for the landscape is based on a simple concept: the fabrication of a floor, walls, roof, or any combination of these elements customized to meet your special needs. Design a unique collaboration of components, engineer them according to structural standards, and you have a structure you can enjoy for many years.

The construction and installation of many of the wooden elements designed to be used in a landscape setting require special skills, tools, designs and a high level of craftsmanship. Some of these elements also face the risk of collapse if the structural elements are not sized properly. Because of this, the construction

Figure 7-1

An open-roofed landscape arbor placed in a garden setting.

of any structure intended to support humans should be contracted to a licensed professional for installation. Structures not intended for occupancy, particularly arbors and trellises, are projects most likely to be safely and successfully completed by the homeowner.

PLANNING THE WORK

You should considers several issues regarding materials when preparing for a carpentry project. The types of wood, fasteners, and finishes need planning and research before you ever pick up a tool. Information regarding the choices of exterior woods and wood-fabrication products is described in the sections following.

Woods and Materials for Exterior Use

Selecting the proper wood for your project requires consideration of several characteristics of wood and how the material is to be used. The primary choices for wood products used in exterior carpentry include naturally decay-resistant and **treated** woods milled into boards and **dimensioned lumber**.

Although treated **softwoods** remain the standard for **structural** (support) components, the landscape market also has a large number of alternatives to harvested woods. Composite and recycled materials are becoming increasingly popular alternatives to woods, for surface trim and finishing. To speed the construction of landscape structures, many components can be purchased prefabricated. Available in a variety of natural and artificial materials, the homeowner can find **lattice**, rail posts, and many other commonly used parts cut and ready for installation. The following sections describe some woods and materials that are commonly used for exterior carpentry.

Western Red Cedar

Cedar heartwood is a soft, naturally decay-resistant lumber that makes excellent surfacing and trim material. Because of its relatively high cost, cedar is usually not cut in structural dimensions (2 × 8's, 2 × 10's, etc.) but is available as a surface and trim lumber. Cedar is soft and can be stained if desired or left to weather out to a mottled gray. **Rough-sawn** cedar boards are often used as trim for exterior work.

Redwood

Redwood heartwood is a soft, naturally decay-resistant lumber that is excellent for decking and trim material. Heart redwood can also be used for structural members in most exterior applications. Redwood is soft and without finishing will weather to a warm dark gray.

Treated Woods

Treated woods include a variety of pines, firs, and spruces that are treated by one of several methods to resist decay and insects. These woods have the strength to be used as structural support components for landscape carpentry but after being treated are not the best or safest choice for surface treatments. See the Cautions regarding wood treatments in the Decay Resistance section on page 166.

Wood composites and recycled materials

Wood-substitute components are available that provide a reliable surface, workability, and some degree of environmental consciousness. These surfaces are typically plastics or a blend of wood fibers and plastic, molded into the shapes and dimensions of deck surfacing. Most of theses **composite** materials are limited to surface treatments and are typically not strong enough for structural components. Sort through the variety of choices available at a local lumberyard, and pick one that offers strength, a fade-resistant color, and ease of workability.

Table 7-1 summarizes the characteristics of hardness, strength (ability to resist breaking when subjected to a load), decay resistance, and common uses for several exterior woods.

Table 7-1 Softwoods

Kind	Hardness	Strength	Decay Resistance	Uses
Red cedar	Soft	Low	Very high	Exterior
Fir	Medium to hard	High	Medium	Framing, millwork, plywood
Ponderosa pine	Medium	Medium	Low	Millwork, trim
Western white pine	Soft to medium	Low	Low	Millwork, trim
Southern yellow pine	Soft to hard	High	Medium	Framing, plywood
Redwood	Soft	Low	Very high	Exterior
Spruce	Medium	Medium	Low	Siding, subflooring

Table 7-1   *Softwoods used in landscape construction.*

Decay Resistance

Woods such as cedar and redwood are naturally resistant to decay, and as long as the appropriate portion of the tree is used, there is no need to further treat these lumbers. In redwood, the **heartwood** portion is used for exterior construction. Most other woods available to the homeowner, including pine, fir, and spruce, are susceptible to decay and insect damage if not treated. Treatment methods for woods include the chemicals pentachloraphenol (also termed Penta), ACQ^R (alkaline copper and quaternary), arsenicals (derived from different copper and zinc compounds), and creosote.

When improperly used, each of these chemicals has the potential to be toxic. Arsenicals are scheduled for discontinued use and creosote is not appropriate in situations in which it comes into contact with humans and plants consumed by humans. It is recommended that structures that bring humans, especially children, into contact with treated lumber be avoided. In these situations, use alternative decay-resistant materials.

These treatments are generally applied to lumber under pressure to force the preservative deeper into the wood. Look for treated woods labeled "Ground Contact" to ensure satisfactory performance in exterior structural projects. When treated wood is cut for use in a project, **retreatment** of the cut ends is necessary. Selection of a paintable or dippable treatment method will be the most efficient method for retreating. After cutting, dip or paint the preservative on the exposed surfaces and let them dry before placing them in contact with the ground.

CAUTION

- Avoid prolonged skin contact with treated lumbers. Working with treated lumbers requires wearing gloves and long-sleeved shirts to prevent exposing your skin to the chemicals used in the treatment.

- Wear a dust mask when cutting treated lumbers.

- Wash your hands thoroughly after contact with treated lumbers, wash the clothes worn when handling treated lumbers separately form other clothes, to prevent contamination.

- Do not let treated lumbers come in contact with food.

- Treated lumber should not be burned and must be disposed of in the manner prescribed by the manufacturer.

Wood Strength

Most structural-dimensioned treated woods (2 × 8's, 2 × 10's, 4 × 4's, and larger) are durable and strong enough to use for posts, beams, joists, and surfaces that must support weight. Smaller-dimensioned lumber (2 × 4's, 2 × 6's) milled from woods such as redwood and cedar and composite materials are best suited for the trim and surfacing elements of the project, where the weight they carry is limited. If you are unsure of the ability of your design to support the weight that will be placed upon it, consult an engineer for advice.

Wood Color and Exterior Finishes

A key esthetic characteristic of any wood product is color. Generally, color can be described in two ways: as the initial color of the lumber and as the ultimate weathered color. Exterior woods have distinctive initial colors, and most will weather to a shade of gray if left unfinished. Cedar is initially a light tan and redwood a reddish brown. The consistency of the final color will depend on whether the surface is exposed to moisture, such as rain, snow, or irrigation, and whether it is under the cover of a roof. The more protection provided, the more consistent the final gray color will be.

Most treated woods begin with either a light green or brown cast. Because the **structural lumber** is not visible when in its final position, the color of treated lumber is not an issue for these pieces; however, if placed in a finish position (that is, where it will be visible when the project is complete), the coloring could be considered unattractive. If you choose to use a treated wood for a finish element, you will have to accept the colored cast left by many of the treatment methods. This discoloration can remain present for several years, but eventually the color will weather. Finishing with stain is an alternative in some cases, but many of the treatments prevent finishes from binding evenly with the wood.

Wood used in exterior conditions will face a variety of weathering conditions that will age the materials, change its color, and possibly begin the process of decay. You should decide during the planning phase of the project whether the wood selected for construction will be treated to deter the effects of weathering or allowed to age naturally. The primary reason to finish-treat such woods is to maintain the new look of the lumber or change the wood's natural color. Once finish-treating is selected for exterior woods it should continue throughout the life of the project.

To preserve the natural look of your wood, a water repellent with an ultraviolet light inhibitor must be applied to slow the weathering process. To change the natural color of the wood, a surface must be stained or painted. Two different types of stains commonly used with exterior applications include opaque stains, which provide a solid color coating over the wood, and transparent stains, which color the surface but allow the grain of the underlying wood to show through.

Cost and Availability

Selection of your wood products will often be dictated by which materials are available at a reasonable cost. Factors such as markets, transportation costs, and production schedules will all influence the price and may necessitate using whatever acceptable material is available. The factors of cost and availability are typically interrelated. The more available a product, the lower the cost. Products that are in short supply are more expensive.

Many of the woods that are treated, such as southern yellow pine and fir, are harvested from large stands and are typically available in greater quantities than woods such as redwood and cedar. Dwindling supplies of redwood have also introduced the ecological issue into material selection, causing some to select alternative choices.

Wood Quality

The quality of the products you use will have a direct relationship to your project's appearance and stability. Avoiding wood that is warped, split, cracked, or has large **knots** will also improve the structural integrity of a project. Some defects, such as cupping, can be overcome by reorienting the lumber if the defect is not severe. Others will require trimming the defective portion from the lumber to make it usable. A serious defect would be cause for rejection of a piece of lumber (Figure 7-2).

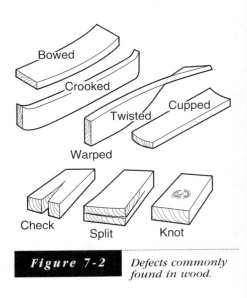

Figure 7-2 *Defects commonly found in wood.*

Lumber Dimensioning

Many people are confused by the sizes of lumber. When a 2 × 4 is first cut, the dimensions are actually 2" thick by 4" wide. The 2 × 4 dimension is the nominal dimension and is used when the product is sold. However, after drying and planing, boards are slightly smaller in their actual thickness and width than the nominal dimension. Using the nominal dimensions rather than the actual dimensions when measuring and cutting will result in errors. To avoid this problem, memorize and use the dimensions listed in Table 7.2.

Table 7-2 Nominal and Actual Dimensions for Stick Lumber

Nominal Dimension	Actual Dimension
1 inch	3/4 inch
2 inches	1-1/2 inches
4 inches	3-1/2 inches
6 inches	5-1/2 inches
8 inches	7-1/4 inches
10 inches	9-1/4 inches
12 inches	11-1/4 inches

Example: 2 x 12 actual dimensions are 1-1/2 inches x 11-1/4 inches.

Table 7-2

Nominal and actual dimensions for dimensioned lumber.

Length is not affected by this issue, but you should verify the length of a piece of lumber by measuring it before installation, to avoid problems.

Fasteners and connectors

Connectors, fasteners, and related hardware for exterior projects should be made of materials that resist rust naturally or are treated to resist rust. **Galvanized** materials are made of steel and then dipped in a zinc coating to reduce the chance of rusting when exposed to moisture. Galvanized materials are suitable for use with any type of treated lumber, but may stain wood if the galvanized coating is damaged. **Polymer-coated fasteners** are made of steel that has then been coated with a thin plasticlike coating to resist rust. Polymer-coated fasteners resist rust and are easy to install because of their slippery coating. Though expensive, stainless steel connectors will provide a stain-free finish when used in contact with cedar or redwood. Your selection of fasteners will depend on how and where you expect your project to be used.

Fasteners

Several types of fasteners could be used in your landscape-construction projects. Among the choices are nails, wood screws, lag screws, carriage bolts, an array of premanufactured fasteners, and specialty connectors (Figure 7-3).

- Nails. Nails are pieces of hardened wire of varying diameters that are cut to length and flattened on one end (called the head) to make it easier to drive them into the wood.

- Wood Screws. Screws are similar in length to nails, but the **shanks** are tapered and threaded at one end so that they can be twisted into the materials to form a stronger hold. Screws can be driven with hand or power screwdrivers using slotted-, Phillips-, or square-headed drives. Screws are best installed by drilling **pilot holes**, holes drilled into the wood where the screw is to be installed. Pilot holes should be the same diameter as the shank of the screw and bored into the wood to a depth equal to at least half the length of the screw. Pilot holes will make installation easier and reduce the possibility of splitting the lumber.

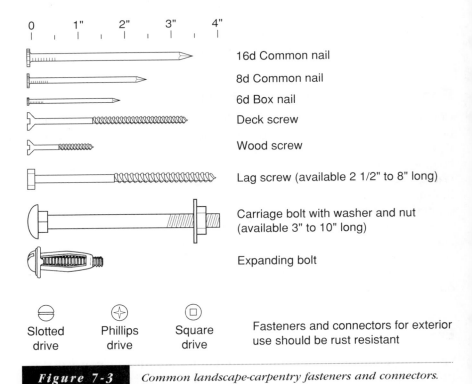

Figure 7-3 Common landscape-carpentry fasteners and connectors.

- Lag Screws. The lag screw is a large-diameter long screw used for anchoring structural members. Lag screws require pilot holes that are the same diameter as the shank of the screw and drilled to a depth of at least two-thirds the length of the screw.

- Carriage Bolts. Carriage bolts are large-diameter bolts with rounded heads and square shoulders and are used for connecting structural pieces. Carriage bolts require a nut and washer at the end opposite the head to complete the connection. Carriage bolts also require that a pilot hole the same diameter as the bolt be drilled through all pieces of lumber being connected.

Specialty Connectors

To improve and speed up the connection of building materials, a wide selection of specialized connecting and support hardware is available. Saddle connectors are available for joining 4 × 4 lumber. Straps, T's, angle braces, and post caps can all be used to make connections that are difficult to nail, screw, or bolt (Figure 7-4). Specialized hardware should be galvanized if it is for exterior use.

Supporting Wooden Structures

Your landscape structure will require some form of support to pre-vent the wind from lifting or blowing the structure over and, in areas in which the ground freezes, anchoring that provides adequate protec-tion from frost heaving. You can do this by supporting your structure on existing slabs, footings or foundations, or posts buried in the ground. More complex methods of support, including anchoring your project to an existing structure, can be accomplished with the aid of a contractor.

Attachment to an Existing Slab or Footing

Connecting your structure to an existing slab will require either drilling and setting an expansion anchor bolt and galvanized-post anchor or making a connection using an angle brace. In cases in which wind may lift the structure, consider using both techniques. Setting an anchor bolt requires marking the center line location of each post, and then using a **masonry bit** to drill a hole the same width and depth as the expansion bolt. Drive the anchor bolt into the hole. Tighten the nut with a wrench

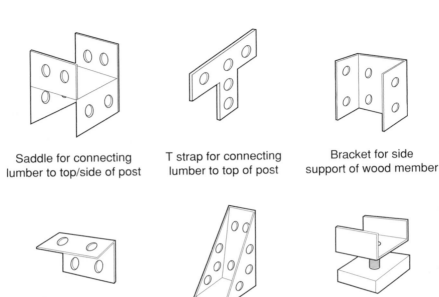

Saddle for connecting lumber to top/side of post	T strap for connecting lumber to top of post	Bracket for side support of wood member
Angle brace for support under wood member	Joist hanger, used for joists and stringers	Post support

Figure 7-4 *Specialty connectors used in landscaping.*

until the bolt fits securely in the concrete (Figure 7-5). Connect a galvanized post anchor to the top of the expansion bolt, and then set the post or beam in the post anchor and tighten the nut.

Angle braces are fastened to concrete using expansion bolts, then lag screwed to a post or beam. Set the post in position and mark the location of the brace on the concrete surface. Drill pilot holes in the concrete the same depth and diameter as the expansion anchor bolt. Install expanding anchor bolts in the concrete and tighten securely with a wrench (Figure 7-5). Place the angle brace over the expansion bolts and tighten the washers and nuts.

Cast-in-Place Concrete Footings

Cast-in-place concrete footings provide a stable support for most types of structures. Structures are anchored to this footing using a galvanized post anchor that is set in the top of the footing while the concrete is still wet.

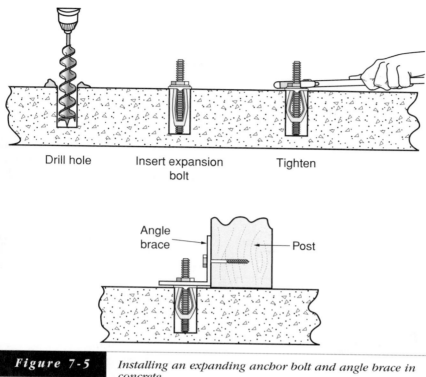

Figure 7-5 *Installing an expanding anchor bolt and angle brace in concrete.*

Drill a pilot hole, insert the expanding bolt, and tighten. The brace should be fastened first to the post and then to the expanding bolt.

Footing anchoring may allow the structure to twist and possibly collapse, so structures that have only posts for support will not work with this type of footing unless the structure is braced (Figure 7-6).

To construct a concrete footing, excavate a 12" diameter round hole to **frost depth** plus 6", centered on the footing-location marking. The hole should be as vertical as possible, with straight sides. A properly excavated hole does not require forms, but if you desire a neat appearance at grade level, a piece of **paper form** or a 12" × 12" square box constructed from 2 × 6's may be used to form the top 6" of the pour (Figure 7-7). Before pouring, recheck your measurements to verify that the footing is located in the proper position and elevation. If it is not, enlarge the hole or fill it and dig another in the proper location.

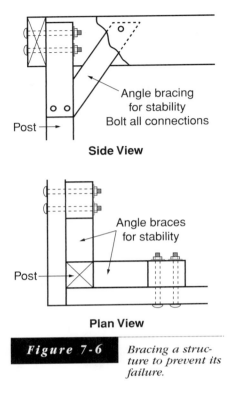

Angle bracing
for stability
Bolt all connections

Side View

Angle braces
for stability

Post

Plan View

Figure 7-6 *Bracing a structure to prevent its failure.*

Reinforce the footing by cutting two #4 (1/2" diameter) **reinforcing rods** (rerod or rebar) 8" longer than the depth of the hole. Insert the reinforcing rod into the hole, pushing the extra length into the ground at the bottom of the hole until the rerods are below the top of footing (Figure 7-7). Place 2–3" of 1" crushed stone into the bottom of the hole for drainage. Mix and pour concrete into the hole and tamp gently with a stick or a shovel handle until it is settled, to work out any air bubbles that may be present below the surface. Fill to the top of the hole and smooth the top with a magnesium float. Slightly dome the top surface of the concrete to enable water to drain off it. Before your footing hardens, insert a galvanized post anchor into the center of the footing; be sure it remains vertical as the concrete hardens. After 24-hours any forming material may be removed from your footing.

If posts are placed on footings or piers, they should be braced and held in correct position until the structure will support itself. To do this, fasten two 1 × 6's, set at right angles to each other, to a post (Figure 7-8). Connect each 1 × 6 to a stake anchored in the ground. Each subsequent post can be braced in two directions in a similar manner or braced against posts previously anchored. Be sure that your project is square before placing any other components.

Direct Burial of Posts

For residential applications in locations with stable subsoils, you can bury treated support posts directly into the ground to provide adequate support for your structure. This type of support provides

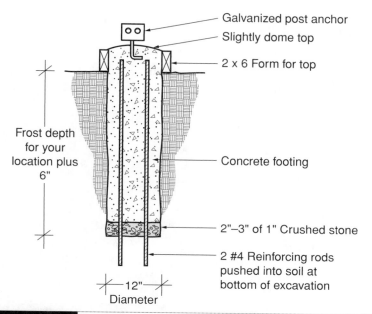

Galvanized post anchor

Slightly dome top

2 x 6 Form for top

Frost depth for your location plus 6"

Concrete footing

2"–3" of 1" Crushed stone

2 #4 Reinforcing rods pushed into soil at bottom of excavation

12" Diameter

Figure 7-7   *Cast-in-place concrete footing.*

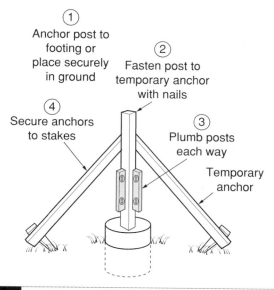

(1) Anchor post to footing or place securely in ground

(2) Fasten post to temporary anchor with nails

(4) Secure anchors to stakes

(3) Plumb posts each way

Temporary anchor

Figure 7-8   *Temporary bracing for posts.*

Wooden Structures ■ *175*

structural support, braces against twisting and, if buried to the proper depth, can provide protection against frost heaving. For direct burial of a post, excavate an 8" diameter hole 6" below maximum frost depth and centered directly on the footing location marking. The hole should be as vertical as possible, with straight sides. Place 2–3" of gravel at the bottom of the hole for drainage. To extend the life of a direct-burial post, wrap the post with asphalt **building paper** at the ground line to prevent contact with soil. The building paper should extend 6" above and 9" below finish grade, then staple it to the post (Figure 7-9).

Figure 7-9

Stapling building paper around a direct-burial post. The paper extends 6" above and 9" below the ground line.

Place the post in the hole, backfill halfway, tamp, and check with a level to ensure that your post is plumb and square. You can backfill with soil excavated from the hole, and the compressed soil will provide stable support of the post if the hole has not been dug too wide. Compact the soil with the handle of a shovel after every 8" of fill has been placed. Another choice for filling around a post is to place concrete in the void around the post. Although this method provides a very stable installation, it may trap moisture around the post, leading to premature decay.

Whichever choice of backfill you select, constantly check that the post is plumb as the hole is filled. Correct leaning posts by wiggling the post and applying pressure in the direction the post needs to move. Recompact after alignment is corrected, and backfill to the finish grade. If you select concrete for backfill, smooth the concrete surface of the hole with a wood float, sloping the surface away from the post so that water will drain away from the post; this will help to prevent rotting of the post.

BUILDING WOODEN STRUCTURES

After making all the preliminary choices and decisions, the actual building of a wood structure may seem anticlimactic. However, if all the work has been properly done, the construction should proceed with minimal problems, and the rewards will last for many years. Two structures suggested for the homeowner to consider are arbors and trellises.

Arbors

Arbors are open-roofed structures that are typically used for shade, to support vining plant material, or to accent entry points in the landscape. Composed of posts that support a framework of open roofing materials, arbors are one of the simplest ways to introduce an overhead enclosure into the landscape. The roof structure is not intended to be weatherproof and can be composed of any number of creative patterns or materials. Vining plant materials are a natural enhancement of the arbor landscape structure, and designs may have the arbor arcing over a walkway or enclosed on one or both sides along a walkway. When designed large enough to cover an outdoor use area, such as a deck or patio, arbors are sometimes termed shade structures.

Construction of a 4' × 8' overhead arbor

Alternatives to this project: Arbors can be designed in several different forms and sizes. See Appendix A for alternative arbor sizes and designs and additional construction details. Alternative arbor 2 (Figures A-1 and A-2) is 8' × 8' with a lattice surface, and alternative arbor 3 (Figures A-3 and A-4) is 10' × 10' with an alternating 2 × 4 and 2 × 2 surface.

CAUTION

■ The components of this project are sized for the design shown. Altering the size will require adjusting the size of the components to maintain structural integrity.

continued

- Verify that the footing depth is appropriate for your geographic region; if you are in an area in which the ground freezes during the winter, you may have to increase the depth of the footing so that it will extend below the frost line.
- Use caution when operating power equipment.
- Follow the manufacturer's instructions when using power equipment.
- Use caution when climbing or working on a ladder.

Time: 1–2 days.

Level: Challenging (17 steps). Heavy lifting and carpentry skills required. Portions of the project will require two people.

Tools Needed:

1. Plan of the project (Figures 7-10 and 7-11).

2. Marking paint.

3. Post-hole excavator (auger or clamshell).

4. Round-nosed shovel.

5. Wheelbarrow.

6. 25' measuring tape.

7. Claw hammer.

8. Carpenter's saw.

9. Pencil.

10. **Circular saw**.

11. Electrical cords and connection to a GFCI power source.

12. Drill (cordless operation or electric) and drill bits (spade or auger, sizes from 1/8" to 1/2" and standard/phillips head screwdriver bits)

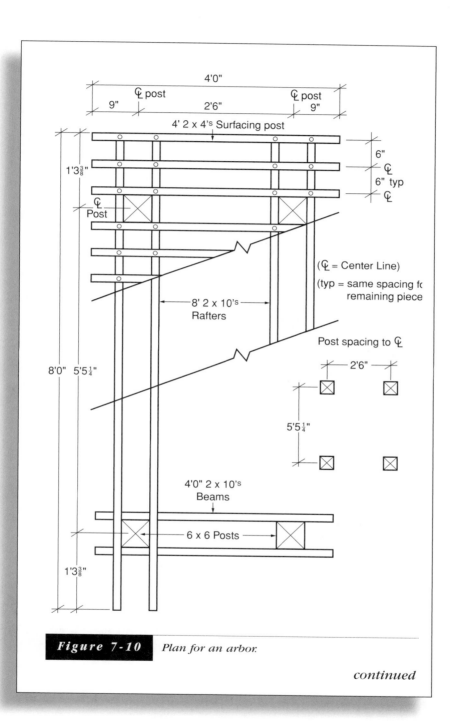

4'0"

ℂ post

9"   2'6"   ℂ post   9"

4' 2 x 4's Surfacing post

1'3⅜"

6"
ℂ
6" typ
ℂ

ℂ
Post

(ℂ = Center Line)

(typ = same spacing fo
        remaining piece

8' 2 x 10's
Rafters

Post spacing to ℂ

2'6"

8'0"   5'5¼"

5'5¼"

4'0" 2 x 10's
Beams

6 x 6 Posts

1'3⅜"

Figure 7-10   *Plan for an arbor.*

continued

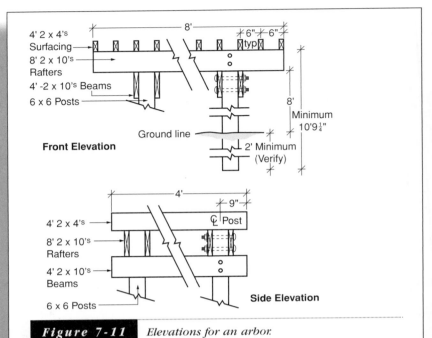

Figure 7-11   *Elevations for an arbor.*

13. Wrench set.

14. 10' step ladder.

15. Carpenter's square.

16. Torpedo level.

17. Stapler.

18. Utility knife.

Materials Needed (based on design shown, with lumber type and finish to be chosen by the builder):

1. Four 6 × 6 posts 12' long (length may be adjusted downward based on burial depth required for your geographic region). (Note: 4" × 4" posts may be substituted, with spacing adjustments, for other wooden structural members).

2. Four 2 × 10 beams 4'-0" long.

3. Four 2 × 10 rafters 8' long.

4. Seventeen 2 × 4 surfacing boards 4' long.

5. Sixteen 3/8" diameter × 10" long carriage bolts with washers and nuts.

6. Seventy #8 diameter × 5" long lag screws.

7. Fifty 16d (16-penny) nails or polymer-coated deck screws.

8. 5 cubic feet (CF) of 1" crushed stone.

9. Finish (stain or paint) selected by the builder.

10. Four 2" × 2" × 2' wood stakes.

11. Four additional 5' and 10' 1 × 4's to use as temporary bracing.

12. Four 15" wide × 15" long strips of building paper. .

13. One box 1/4" staples.

Directions:

1. If you plan to directly bury your support posts, attach a strip of building felt to each, as described in the previous section titled Direct Burial of Posts.

2. Find post locations on the plan (Figure 7-10). Paint a large × mark on the ground, with the center of the mark indicating where the center of the post will be.

3. Excavate each post location to frost depth for your area or to a minimum of 24" deep.

4. Place 1/2" CF of crushed stone in the base of the each hole.

5. Set one post in each hole.

6. Hold posts in correct position and, about 4' above grade, nail the 2 × 4 braces between the posts. Backfill the holes. Check the posts for plumb and proper spacing often (Figure 7-12). When completely backfilled, the posts should be plumb and spaced according to the plan.

continued

7. With posts secure, temporarily nail the 4'- 0" 2 × 10 beams at the height shown in the front and side elevation drawings (Figure 7-11).

8. Drill holes through the 2 × 10 beams and posts at the locations shown. Install carriage bolts through the pilot holes and secure with washers and nuts. Repeat for all four posts. Remove the temporary nails Figure 7-13).

9. Position the 8' 2 × 10 rafters on top of the 2 × 10 beams as shown in the drawing and temporarily nail in position. Check the structure for square (that is, diagonal post-to-post measurements are equal) and level.

Figure 7-12 *Setting and leveling posts.*

10. Drill holes through the 8' 2 × 10 and posts at the locations shown. Install carriage bolts through the pilot holes and secure with washers and nuts. Repeat for all four posts and remove the temporary nails.

11. With the carpenter's saw, trim the top of the posts flush with the tops of the 8' 2 × 10's. If you are using treated posts, retreat the cut ends.

12. Remove any temporary bracing.

13. Place the end 4' 2 × 4 surfacing on its edge on top of the 8' 2 × 10's in the location shown on the drawing.

14. At each location where the 2 × 4 crosses the 2 × 10, drill a pilot hole three-quarters of the way through the 2 × 4. (Note: it may be easier to mark one 2 × 4 to use as a pattern for the rest, then mark and drill the holes on the ground rather than on top of the structure.)

15. Insert a #8 × 5" lag screw through each pilot hole and twist it into the 2 × 10 using a wrench.

16. Repeat this installation of 2 × 4's for each location shown on the drawing (Figure 7-14).

17. Apply the finish you chose.

Figure 7-13

Post connection using carriage bolts. At this corner bolts are installed from both sides.

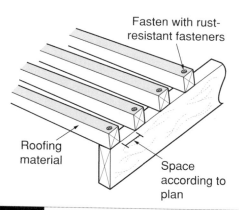

Fasten with rust-resistant fasteners

Roofing material

Space according to plan

Figure 7-14   *Open roof connections.*

Trellises

Trellises are vertical structures with open frameworks, often with vining plant growth growing upon them. Trellises may be free standing or can be attached to walls, arbor posts, or the sides of structures. If creatively used, trellises can function as both a screen to hide undesirable areas and as a support for growing fruiting and flowering plants (Figure 7-15). When considering trellis designs, choose one that will have sufficient strength to support the weight of aggressive vining plants. People often don't recognize that the weight and growth habit of a mature vining plant is enough to cause a trellis to collapse.

Figure 7-15 *Landscape trellis supporting grapevines.*

Construction of a trellis

Alternative to this project: Trellises can be designed in a variety of forms. See Appendix B for an alternative trellis. The alternative trellis (Figures 11-7 and 11-8), is a post-and-cable trellis that employs turnbuckles to keep the framework taut.

CAUTION

- Use caution when operating power equipment.
- Follow the manufacturer's instructions on the use of power equipment.
- Use caution when climbing on ladders.

Time: 1–2 days.

Level: Challenging (14 steps). Requiring physical exertion and carpentry skills. Portions of the project will require two people.

Tools Needed:

1. Plan of project (Figure 7-16).

2. Marking paint.

3. Post-hole excavator.

4. Round-nosed shovel.

5. Wheelbarrow.

6. 25' tape measure.

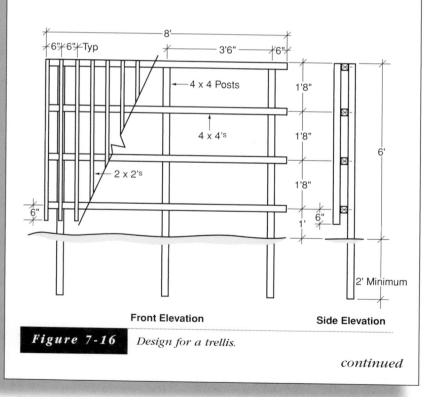

Front Elevation　　　**Side Elevation**

Figure 7-16　*Design for a trellis.*

continued

7. Claw hammer.

8. Carpenter's saw.

9. Pencil.

10. Circular saw.

11. Power cords and connection to GFCI power source.

12. Drill (cordless operation or electric) and drill bits (spade or auger, sized from 1/8" to 1/2", standard and phillips head screwdriver bits).

13. Wrench set.

14. 10' step ladder.

15. Carpenter's square.

16. Torpedo level.

17. Stapler.

18. Utility knife.

Materials Needed (based on design shown; lumber type and finish to be chosen by builder):

1. Three 4×4 posts 12' long (length may be adjusted downward based on burial depth required for your geographic region)

2. Four 4×4's 8' long.

3. Nineteen 2×2's 5' long.

4. Twelve 3/8" diameter $\times$ 6" long carriage bolts with washers and nuts.

5. Fifty 3" long polymer coated deck (or wood) screws.

6. Five cubic feet of 1" crushed stone.

7. Finish selected by builder.

8. Four additional 10' 1 × 4's to use as temporary bracing.

9. Four 2" × 2" × 2' wood stakes.

10. Fifty 16d nails.

11. Four 15" wide × 15" long strips of building paper.

12. One box of 1/4" staples.

Directions:

1. If directly buried, attach a strip of building felt as described in the earlier section titled Direct Burial of Posts.

2. Determine post locations from the plan. Paint a large × mark on the ground, with the center of the mark indicating where the center of the post will be.

3. Excavate each post location to at least 6" below maximum frost depth for your area (a minimum of 24" deep).

4. Place 1/2" cubic foot of crushed stone in the bottom of each hole.

5. Set one post in each hole.

6. Hold posts in correct position and, about 4' above grade, nail a 2 × 4 between each of the posts as temporary braces. Backfill the holes. Check posts for plumb and proper spacing often. When completely backfilled, the posts should be in the position shown on the plan.

7. With posts secure, have an assistant hold the 8' 4 × 4's at the height shown in the drawing.

8. Drill holes through the 4 × 4's and posts at the locations shown. Install carriage bolts through the pilot holes and secure them with washers and nuts. Repeat at all locations where 4 × 4's intersect posts.

9. Remove the temporary bracing.

continued

10. Place the edge 5' 2 × 2' vertically in front of the 8' 4 × 4's in the location shown on the drawing.

11. At each location where the 2 × 2 crosses the 4 × 4, drill a pilot hole, slightly smaller than the diameter of the fastener, through the 2 × 2.

12. Insert a 3" wood screw through each pilot hole and drive it into the 2 × 10 using the drill.

13. Repeat this installation of 2 × 2's for each location shown on the drawing.

14. Apply your chosen finish.

Chapter 8

FENCES AND FREE-STANDING WALLS

If your site needs a screen, enclosure, or boundary definition, you will most likely choose a fence or free-standing wall to fulfill this function. Fencing and walls have been used for thousands of years to create privacy and to separate one property from another, but landscape design has advanced beyond obvious functional choices. In addition to performing the basic role of defining space, fences and walls are now also used to make dramatic design statements. Intricate patterns, creative materials, and attention to detail can add richness to a design. Boundary definition has moved beyond the split–rail fence, leaving your imagination as the only barrier to artistically defining a space.

To fulfill the function of using natural materials to create space and define boundaries, the stone wall has few equals. Stone assembled as a **free-standing wall** creates one of the most lasting impressions of hand craftsmanship. Walls are considered free-standing if their primary use is to define space rather than to retain the soil of an embankment. Although all walls require special skills to build, construction of a free-standing wall requires equal parts technical skill and artistry, in order to assemble a stable and attractive structure.

Wooden fencing offers the widest variety of options, with wood-surfaced fences and prefabricated panel fences offering a wide range of visual appearances (Figure 8-1). Wood panels can be purchased or custom built to create an almost infinite range of patterns. The earthy character of wooden fencing is a standard in creating and separating space. Fencing options in vinyl provide an easy-to-assemble, low-maintenance option for perimeter definition that mimics wood. **Chain link** provides a durable, secure enclosure for pets, play yards, and property. Although vinyl and

Figure 8-1   *Wooden fence.*

chain link will seldom win design awards, each offers characteristics that may make them important to the homeowner. Regardless of your choice, any material can serve for years if properly planned and installed.

The variety of fence and wall types available creates a world of options, ranging from attractive to functional for the homeowner. Fences of decorative metal set in footings should be left to the qualified contractor. Despite this restriction, however, many options remain within the capabilities of the homeowner. Choices for your fence could include wood, chain link, or vinyl. Similar options exist with walls. Dry-laid stone walls can be built by the homeowner, but walls that require forming and pouring concrete, stone with mortared joints, and masonry units are best installed by a qualified contractor.

PLANNING THE PROJECT

As with all projects, time spent planning will reduce the headaches of correcting problems, and fences and walls are no exception. Several factors should be considered before beginning to construct a fence or wall, the most important of which is making certain that the project is located on your property. If your project is along such a boundary, ascertaining the exact location of the property line between neighbors is a task that is best left to the **land surveyor** for accuracy and liability reasons. Although it is expensive to hire a professional for this task, the cost of correcting mistakes and dealing with legal boundary issues is much higher.

For any fencing work, obtain permission from adjacent landowners if the work requires traversing their property, and always leave their property clean and in good condition. In addition to locating the fence or wall, consider choices in patterns, layout, maintenance, and related issues before beginning the project. Local regulations may also alter your plans. Restrictions on locations, heights, and materials are often part of zoning regulations.

Patterns

Fence design is limited only by the designer's creativity and the properties of the materials used. Numerous possibilities exist for all fencing materials, with the ultimate choice balancing appearance, cost, ease of installation, and maintenance. The design of your fence should also consider all views of the fence; maintaining good relations with neighbors often requires selecting a fence pattern that looks good from both sides. Patterns can be copied from fences you have found or seen in magazines or from examples shown in Figure 8-12, with details in Appendix C, Figures C-9 and C-10.

Fence Alignment and Wind Resistance

Fences have significant surface areas that can present substantial wind resistance, which in turn can result in the fence being damaged or even collapsing. Adding variations in alignment to a fence increases its resistance to wind (Figure 8-2). Consideration should be given to designing the fence with **niches,** or corners, and avoiding long straight runs of fence. One way to strengthen the fence and neighborhood relations is to build a niche which provides a doorway to the neighbors or a cove for them to install a favorite plant. While building the fence, temporary bracing should be placed for support.

Layout

Begin the fence project by installing a stringline at the fence alignment. Locate all posts along this alignment using flags or paint markings. Posts should be marked center to center. Locate any slopes where post pattern or length may have to be adjusted to step up the grade. Once end posts have been set, a stringline can be run from one end post to the

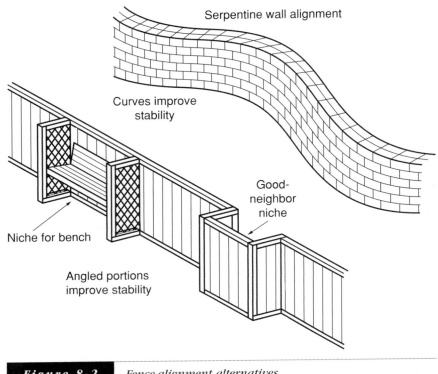

Serpentine wall alignment

Curves improve stability

Good-neighbor niche

Niche for bench

Angled portions improve stability

Figure 8-2 *Fence-alignment alternatives.*

other on each leg of the fence. This will provide a guide for setting the depth of posts between corners.

Fencing on Slopes

Most fencing and walls can follow moderate slopes without losing function and aesthetic appeal, but special planning and construction techniques are required when traversing hills. Wood-panel fencing, constructed with panels suspended between each pair of posts, typically "steps" along slopes of all degrees. To cover slopes, the posts are extended upward and the panels shifted up or down to match the grade Figure 8-3).

Maintenance Strips for Fences

Maintenance around fence installations can be a time-consuming chore for the homeowner or grounds keeper. When installing fences,

consider installing a maintenance edge under its length to
reduce the time and effort
required to trim and weed. The
maintenance edge can be a 4"
thick × 1' wide strip of concrete or stone strip over weed
barrier. After the fence alignment is determined, remove the
sod along this strip. Install the
paving materials after the posts
have been set and before the
fence surfacing is installed
(Figure 8-4).

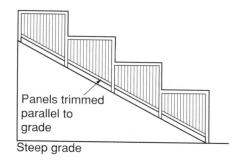

Panels trimmed
parallel to
grade

Steep grade

Panels at angle to ground

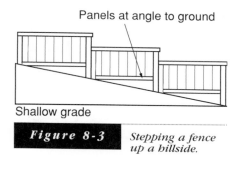

Shallow grade

Figure 8-3   *Stepping a fence
up a hillside.*

FENCE AND WALL INSTALLATION

Choosing between a wall and a
fence is not an easy process.
One consideration in guiding
your choice is the setting in
which you are building: Urban
situations and small lots often
place space at a premium,
making the fence a logical
choice. Conversely, open areas
and large lots invoke images of
stone walls used to separate
fields. Materials used in local
architecture can also help you
make the choice, with buildings
and paving of stone best com-
plemented by repeating those
materials. Wood, brick, and concrete blend well with wooden
fences.

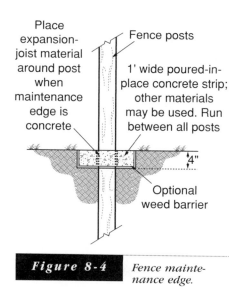

Place
expansion-
joist material
around post
when
maintenance
edge is
concrete

Fence posts

1' wide poured-in-
place concrete strip;
other materials
may be used. Run
between all posts

4"

Optional
weed barrier

Figure 8-4   *Fence maintenance edge.*

Wood-Surfaced Stringer Fencing

Wood-surfaced stringer fencing, also call wood-surfaced fencing, is an attractive and functional boundary definer. The term stringer fencing comes from the horizontal structural pieces installed in the fence that support the wood surfacing. Unlike panel fences, wood stringer fencing is custom built on the site. When compared to other fence types, wood-surfaced stringer fences may require additional time to install because of the one-piece-at-a-time fabrication required. To improve the appearance, trim is typically added to the post surface and stringers to hide structural components.

Figure 8-5 *Angles used to support a fence stringer.*

A basic component of your wood-surfaced stringer fence is the 4×4 posts, which are spaced evenly along the length of the fence. The spacing of the posts will create segments of fence that typically should not exceed 8' in length and may be spaced closer if navigating a grade or improved structural support is desired. Posts should be buried to the frost depth in your geographic area, or a minimum of 24".

Stringers, typically 4×4's, are attached at the top and bottom between each pair of posts. You can attach these easily using specialized hardware called angles (Figure 8-5). Fences over 4' tall or with a two-pattern design may also require a third stringer placed in the center. As an alternative, 2×4 stringers may be attached to the face of the posts, but this option limits the number of surface patterns available. Whatever surface pattern you choose is then attached to the stringers. Trimming the faces of posts and stringers using $1 \times$ lumber and installing gates completes the fence.

Installing a wood-surfaced stringer fence

CAUTION

- Locate all utility lines prior to construction.
- Use caution when cutting materials.

Time: Thirty minutes–1 hour per linear foot (LF) of fence, depending on the complexity of the surface pattern.

Level: Challenging (25 steps). Digging and lifting required. Two people may be needed to complete installation.

Tools Needed:

1. Plan for fence installation. Details for some common surfacing patterns can be found in Appendix C, Figures C-9 and C-10.
2. 25' measuring tape.
3. Marking paint.
4. Marking pen or carpenter's pencil.
5. Chalk line.
6. 100' roll of mason's twine.
7. Carpenter's level.
8. Clamshell or auger post-hole excavator.
9. Round-nosed shovel.
10. Wheelbarrow.
11. Circular saw.
12. Power cords and connection to a GFCI power source.
13. Carpenter's square.
14. Claw hammer.
15. Screwdriver (cordless operation or electrical) and bits.
16. Paintbrush.

continued

Materials Needed:

1. Fencing materials. Calculate the quantities needed, based on the design and length of fence. Materials required will include:

   - Treated 4×4 posts. (One post per fence segment plus one. The length will be determined by frost depth in your geographic region. Extend posts on either side of a gate to 42" below grade.)

   - Treated 4×4 stringers (Minimum two per segment, depending on design. Their length should equal the panel length.) Note: Your design may call for 2×4 stringers.

   - Surfacing and trimming material (purchased based on your choice of surface pattern).

   - Galvanized angle supports for stringers (4 per segment plus 4 extra).

   - Premanufactured gate and hardware (one set for each gate).

2. Galvanized joist-hanger nails.

3. Galvanized or polymer-coated 2" screws (1/2 lb. per panel). Longer screws may be required for some surfacing materials.

4. 8d galvanized box nails (1 lb. per panel).

5. Angular 1" crushed stone (1/2 CF per post).

6. Wood finish as selected by owner.

7. Wood preservative (1 quart for every 50 LF of fence).

Directions:

Post installation:

1. Review the plan for the spacing of posts. Each segment of the fence will have a post at either end with at least two stringers between posts.

2. Mark each post location carefully so segment length will be consistent. If the length of the fence does not allow all segments to be the same length, place one small segment at either end of the fence to accommodate any excess.

3. In the center of every post location, dig or auger an 8" diameter hole 6" deeper than the required burial depth of the post.

4. Using a marker, draw a line near the base of the post that indicates the proper burial depth. To save time with height adjustments, mark and brace corner posts first, then stretch a stringline between the tops of corner posts to use as a guide for remaining posts (Figure 8-6, Steps A through H). If posts are too tall, trim with a circular saw and treat the cut end with a wood preservative.

5. Place 1/2 CF crushed stone in the bottom of each hole and set posts in the holes. Adjust the height of each post by adding aggregate until the mark is at finish grade. Before backfilling, check the spacing between posts to verify that it matches segment dimensions.

6. Place backfill in the hole and tamp with a shovel handle. Recheck for height, plumb, and spacing when the hole is half filled, and completely filled.

7. Install posts on a hillside plumb, and make them extra tall, so that the fence segment can be installed level with the height at least as tall as level segments. Add the amount that the slope drops between posts to the height of each post (Figure 8-7). After installing the panel, trim off the excess post height. Follow these steps for each pair of posts on a slope.

Stringer installation:

1. Mark the locations where the stringers are to be positioned on the post. Stringers should be installed level between the posts, with the top stringer at the top of the posts and the bottom stringer no more than 12" above ground level. If the panel needs additional support, a middle stringer can be positioned midway between the top and bottom or where required by the design.

2. Hold a 4 × 4 up to one of the level lines and mark the distance between the posts. Cut the stringers to this length and treat the cut ends with a wood preservative.

continued

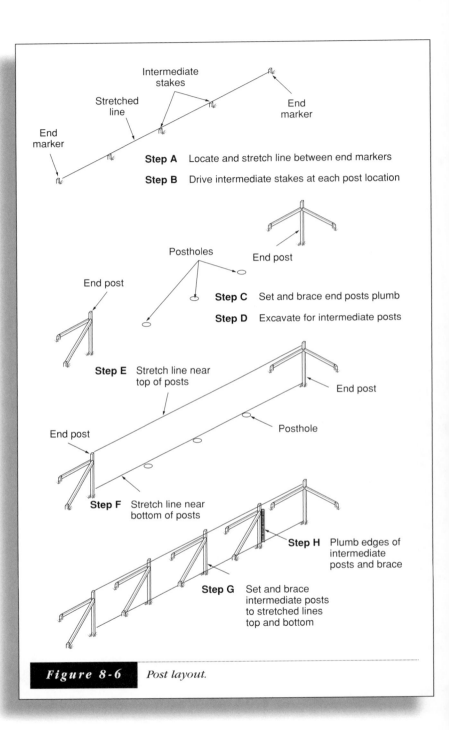

Intermediate
stakes

Stretched
line

End
marker

End
marker

End
marker

Step A Locate and stretch line between end markers

Step B Drive intermediate stakes at each post location

Postholes

End post

End post

Step C Set and brace end posts plumb

Step D Excavate for intermediate posts

Step E Stretch line near
top of posts

End post

End post

Posthole

Step F Stretch line near
bottom of posts

Step H Plumb edges of
intermediate
posts and brace

Step G Set and brace
intermediate posts
to stretched lines
top and bottom

Figure 8-6   *Post layout.*

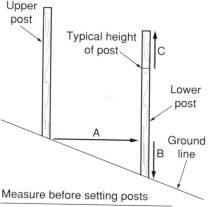

Upper post

Typical height of post

C

Lower post

A

Ground line

B

Measure before setting posts

A Extend level line from ground line on higher post to lower post

B Measure down from level line at shorter post

C Add measurement from B to typical height of lower post

Repeat for each pair of posts

Figure 8-7 *Determining post height for stepped-fence panels or segments.*

3. Install metal hangers at the top and, if present, the middle, stringer marks, making allowance for the thickness of the lumber that will rest on the hanger.

4. Set the stringers on the hangers. Fasten at all locations (Figure 8-8).

5. For the bottom stringer, attach the metal hanger to the stringer before installation. Position the stringer at the correct height and nail the hanger to the post.

Surfacing installation:

1. Measure and cut surface material for the fence. Vertical facing can be extended up to 1' beyond the stringers, but extensions longer than this tend to warp.

continued

2. Place a stringline along the fence run at the height of the top of the fencing material.

3. Attach the surfacing at the top using a 2" screw, and then check for plumb.

4. Finish attaching the surfacing at the top and bottom using two 2" screws per stringer (Figure 8-9). If vertical surfacing is designed with spaces between boards (such as alternating board or picket fences), use a **spacer** to maintain consistent spacing along the entire fence run. Plan the spacing so that a board or picket is placed over the support posts.

5. Alternative vertical surfacing may require different preparation before attaching materials (see Appendix C for fence-surfacing details).

Figure 8-8 *Stringer attachment for vertical surfacing.*

Figure 8-9 *Vertical surfacing attachment.*

- If you are using grape-stake surfacing, drill pilot holes at each location where surfacing will contact a stringer. Use deck screws to attach grape stakes.

- Attach stockade fencing without space between surfacing pieces.

- For vertical louvered fencing, use only a top and bottom stringer. Mark the location and angle of each louver on both the top and bottom stringers, and then attach a block of 2" decay-resistant wood with deck screws at each mark. Trim any portion of this block that extends beyond the stringers. Attach the louvers to the block at the top and bottom with two screws at each location.

- If you are using lattice surfacing, support pieces for the lattice are attached to the inside of the stringers and support posts using 8d galvanized nails. Cut the lattice to fit and fasten it to the support pieces using 6d galvanized nails. Place trim over the lattice to prevent wind from loosening the surfacing. Space horizontal and vertical supports no more than 2' apart in each direction, to provide proper support for lattice.

Trimming:

1. 1 × 6 or 1 × 8 trimming material (dimensions will be determined by the design) can be installed at the top, posts, and corners of a fence to improve its appearance.

2. Measure and cut the trim using bevel cuts and attach using 8d box nails (Figure 8-10).

Gate installation:

1. Measure the gate opening and check it for square. From these dimensions, subtract the hinge and latch dimensions and an additional 1/4" safety margin. Build a gate framework out of 2 × 4 material that is slightly smaller than the gate opening minus the above measurements (Figure 8-11).

2. Add corner bracing or a diagonal brace made of a 2 × 4.

continued

3. Apply surfacing that matches the surfacing used for the fence.

4. Attach the hinges and latch hardware to the gate panel.

5. Prop gate in position in the gate opening. Connect the hinges to one post and check the gate for proper swing.

6. Attach the latch hardware to the other post and adjust for proper closing.

Figure 8-10 *Fence trimming.*

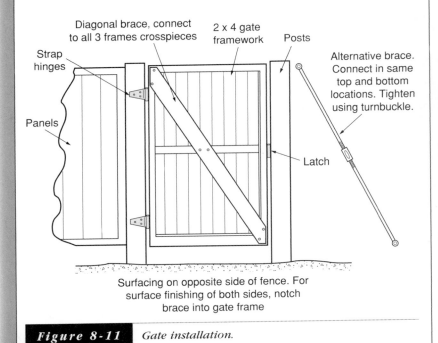

Diagonal brace, connect to all 3 frames crosspieces

2 x 4 gate framework

Posts

Strap hinges

Alternative brace. Connect in same top and bottom locations. Tighten using turnbuckle.

Panels

Latch

Surfacing on opposite side of fence. For surface finishing of both sides, notch brace into gate frame

Figure 8-11 *Gate installation.*

Wooden Prefabricated Panel Fencing

Wooden prefabricated panel fencing can create an attractive perimeter for lawn enclosures and serve as a decorative landscape element. Panel screens can be all of a single pattern or several different patterns (Figure 8-12). Panels can also be custom manufactured to visually coordinate with forms and shapes from the landscape design. Compared to other fencing materials, wood is moderately priced and ranks average in maintenance, requiring periodic refinishing and repairs. Installing it requires basic carpentry skills and good planning. When creatively used, wooden panels can traverse any slope and perform any function expected of any other fencing material.

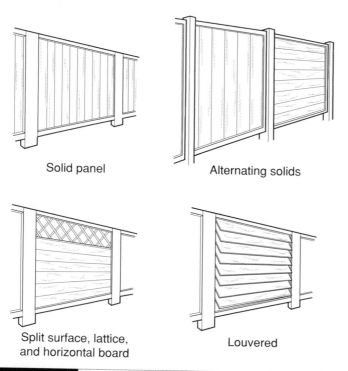

Solid panel

Alternating solids

Split surface, lattice, and horizontal board

Louvered

Figure 8-12   *Common wooden-panel fence patterns.*

Installing a wooden panel fence

CAUTION

- Locate all utility lines prior to construction.
- Use caution when cutting materials.

Time: 30 minutes–1 hour per LF of fence.

Level: Challenging (16 steps). Digging and lifting required.
Two people may be needed to complete installation.

Tools Needed:

1. Plan for fence installation.

2. Marking paint.

3. Marking pen.

4. Chalk line.

5. 25' tape measure.

6. Carpenter's level.

7. Post-hole excavator.

8. Round-nosed shovel.

9. Wheelbarrow.

10. Circular saw.

11. Power cords and connection to a GFCI power source.

12. Claw hammer.

13. Screwdriver (cordless operation or electrical) and bits.

14. Paintbrush.

Materials Needed:

1. Fencing materials. Calculate the quantities needed, based on the design and length of your fence. Materials required will include:

   - ■ Posts, one post per panel plus one. Length will be determined by frost depth in your geographic region. Extend posts on either side of a gate to 42" below grade.

   - ■ Surfacing panels (purchase prepared panels or construct your own design).

   - ■ Specialized hangers for fence panels (4 per panel plus 4 extra).

   - ■ Premanufactured gate and hardware (one set for each gate).

2. Galvanized joist-hanger nails.

3. Galvanized or polymer-coated deck screws (1/2 lb. per panel).

4. 8d galvanized nails (1/2 lb. per panel).

5. Angular 1" crushed stone (1/2 CF per post).

6. Wood finish as selected by owner.

Directions:

Post installation:

1. Review the plan or manufacturer's specifications for spacing the posts. Face attachment requires different post spacing than hanging panels between posts.

2. Mark each post location carefully, to minimize adjustment of panel length. If the length of the fence does not allow all panels to be of equal length, place the smaller panels at either end of the fence.

3. In the center of every post location, dig or auger an 8" diameter hole 6" deeper than the burial depth of the post.

continued

4. Using a marker, draw a line near the base of the post that indicates the proper burial depth. To save time with height adjustments, mark and brace corner posts first, then stretch a stringline between the tops of corner posts to use as a guide for the remaining posts (Figure 8-6, Steps A through H). If posts are too tall, trim them with a circular saw and treat the cut ends with a wood preservative.

5. Place 1/2 CF crushed stone in the bottom of each hole and set posts in the holes. Adjust the height of each post by adding aggregate until the mark is at finish grade. Before backfilling, check the spacing between posts to verify that it matches the panel dimensions.

6. Begin placing backfill in the hole and tamping with a shovel handle. Check for height, plumb, and spacing when the hole is half filled and again when completely filled.

7. Posts on a hillside should also be installed plumb and should have extra height to allow panels to be installed level. To the height of each post, add the amount that the slope drops between ports. (Figure 8-7). After panel installation, trim off the excess post height. Repeat this process for each pair of posts on a slope.

Panel Installation:

Panels for fences are preconstructed and attached between posts using metal hangers (hung panels) or are screwed to the face of the posts (face mounted). For installation methods, hold the panel in position to check for fit. If it is too big, trim a small amount off the end of the panel using a circular saw. If panels are made of treated lumber, treat the cut edge.

Installing hung panels:

1. Mark the top location of all panels using a chalk line. For panels on a hillside, use a carpenter's level to assure that the top is properly aligned.

2. Install the hangers on the sides of the posts using 8d galvanized nails or deck screws.

3. Slide the panels down between the hanger flanges.

4. While holding the panel in the proper position, attach it to the hangers using 8d galvanized nails (Figure 8-13).

Installing face-mounted panels:

1. Mark the top location of all panels using a chalk line. For panels on a hillside, use a carpenter's level to assure that the top is properly aligned.

2. Hold the panels at proper height and attach them to the face of the posts using four deck screws per side (Figure 8-13).

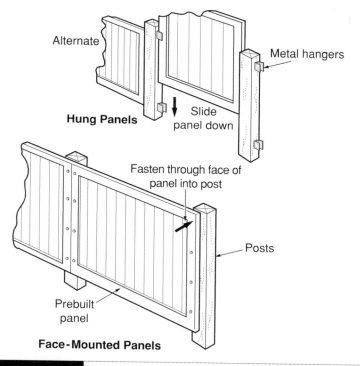

Figure 8-13 *Prefabricated panel fencing: panel installation. Surface mounted panels are shown below with panels hung between posts shown above.*

continued

Gate installation:

Gates are typically premanufactured for panel fences, leaving only the hardware attachment to complete. If the gate is not premanufactured, follow directions for Gate Construction as outlined on page 201.

1. Attach the hinges and latch hardware to the gate panel and prop it in position in the gate opening.

2. Connect the hinges to one post and check the gate for proper swing.

3. Attach the latch hardware to the other post and adjust for proper closing.

Vinyl Fencing

Vinyl fencing is a plastic product available in patterns that mimic wooden-**rail** fencing. Although its costs are moderate, maintenance low, and installation relatively easy, many designers consider vinyl only a secondary option. Limited colors and patterns, and the imposition of a non-natural material into the landscape offend many purists. Those who do use vinyl base their decision on a match with specific design considerations or maintenance needs.

Vinyl is installed in a progressive manner, with postholes being located and excavated, but with posts not set until the railings have been installed. Once the installation is begun, a rhythm will develop for the placing of posts, inserting rails, and securing posts. This series of steps is repeated for each set of posts and rails throughout the run of the fence.

Installing a vinyl fence

CAUTION

- ◼ Locate all utilities prior to construction.
- ◼ Use caution when cutting materials.

Time: 30 minutes–2 hours per LF of fence.

Level: Moderate (23 steps). Digging required. Two people may be needed to complete installation.

Tools Needed:

1. Plan for fence installation.

2. Marking paint.

3. Marking pen.

4. Chalk line.

5. 100' roll of mason's twine.

6. 25' tape measure.

7. Carpenter's level.

8. Clamshell or auger fencepost excavator.

9. Round-nosed shovel.

10. Wheelbarrow.

11. Hacksaw.

12. Claw hammer.

13. Drill and bits (cordless operation or provide power for saw).

14. Screwdriver.

15. Special **crimping** tool (necessary for many fence brands).

continued

Materials Needed:

1. Vinyl fence components, including:

   ■ Posts (number of posts determined by post spacing).

   ■ Rails. (Calculate rail lengths by dividing LF of fence by length of each rail. Multiply the answer by the number of rails in the fence.)

   ■ Caps (one per post).

   ■ Connectors (typically pins designed for the fencing system).

2. Angular 1" crushed stone. (1/2 CF per post)

Directions:

Post Installation:

1. Review the plan for post locations.

2. Using the paint, mark each post location carefully on the ground so that minimal adjustment of segment length will be required. Post spacing is determined by the length of the rails or panels provided. If you can maintain regular spacing, construction will be considerably easier. If regular spacing cannot be maintained, trim the rails or panels and adjust the spacing of posts to create equal segments along the entire fence.

3. Dig or auger an 8" diameter hole to the burial depth of the post at every post-center location.

4. Using a marker, draw a line near the base of the post that indicates the proper burial depth.

5. Special posts may be available for corners and ends of runs. Verify that the proper post has been selected for each location.

6. Insert posts into the holes and check the postholes for proper depth, but do not backfill or anchor posts in holes until after rail has been installed (Figure 8-14). If posts are too tall, they must be removed and the excavation deepened. If

the post sets too low, add granular material to the hole to raise the elevation.

7. Posts for rail fences should be installed plumb and at the same height above ground level. Rails will parallel the slope. Verify correct post spacing by holding one of the rails between an installed post and the next post to be dug.

8. For abrupt changes in slope (those over 2' fall in 100 feet of distance), you may need to cut rail sections and decrease the spacing between posts.

For rail installation, follow the instruction set that relates to your rail length. Long rails connect three or more posts, whereas short rails connect only two posts.

Long-Rail Vinyl Fences.

One form of rail fencing uses rectangular openings in a post through which the rail section can be inserted.

1. Set the first three posts loose (use a corner post if beginning at a corner) in their holes and adjust to their correct height.

2. Using the crimping tool, crimp one end of a rail on at least two sides (Figure 8-15). Crimping deforms the surface of the rail to prevent it from sliding out once it

Figure 8-14

Vinyl fencing: post installation. Post to the left is set with rails installed. Two posts to the right are loose to allow rail installation and adjustment.

continued

has been inserted into a post. The rail is not crimped at the middle post. Slide the uncrimped end of the rail through the bottom hole of the middle post and push it through until the rail is even with the first post (Figure 8-16, Step A).

3. Insert the crimped end of the rail into the bottom opening of the first post (Figure 8-16, Step B).

4. Repeat this crimping-and-rail-installation process with all rails in this section.

5. When all rails are installed into the first and through the middle posts, check the length of the rails against the third post in the section. If the rails, after inserted, will extend beyond the center of the third post, trim them with a hacksaw. Then crimp the unattached end of the rail and slide them into the openings in the third post (Figure 8-16, Step C).

6. After all rails have been installed, set the posts at the level indicated by the depth mark on each post. Verify that the rails are parallel to the ground and that the posts are set along the alignment desired. Adjust post height if necessary and backfill around the posts.

7. Repeat this process for the next rail section, using the third post in the previous segment as the first post for the next set.

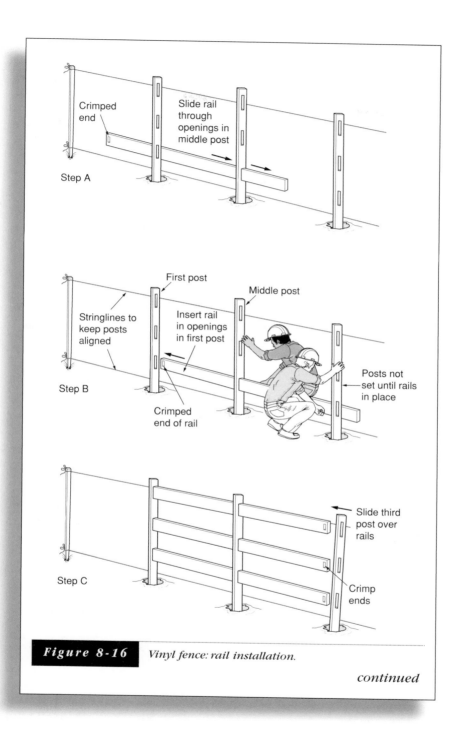

Figure 8-16   *Vinyl fence: rail installation.*

continued

Short-Rail Vinyl Fences:

Short rail fences use short railings placed between a pair of posts.

1. Set the first two posts (use a corner post if beginning at a corner) in their holes. The posts should be loose in the hole when you are performing rail installation.

2. Crimp at least two sides of both ends of the rails for one section. If pins and predrilled holes are used, crimping will not be necessary. Install pins after each rail end has been placed in the post.

3. Place one end of the bottom rail in the first post.

4. Place the other end of the bottom rail in the second post. Fasten with pins if required. Installation is easier if the top of the second post is angled slightly away from the first post, and then gradually pulled to a vertical position as each rail is installed. The rail should not extend beyond the center of the post into which it has been inserted. If minor length adjustments are required, trim the rail with a hacksaw.

5. Insert the next rail in the first post and then in the second post. Fasten with pins if required.

6. Place all remaining rails in the first post and then in the second post, and pull the second post towards a vertical position. Complete fastening with pins if required.

7. After all rails have been installed, set the posts at the level indicated by the depth mark on each post. Verify that the rails are parallel to the ground and that the posts are set along the alignment desired. Adjust post height if necessary and backfill around both posts.

8. Repeat this process for the next rail section, using the second post in the previous segment as the first post for the next set.

Chain-Link Fencing

Chain-link fencing is composed of metal posts with a woven wire fabric stretched between the posts. Although most applications of chain link are for purposes of security or boundary protection, colored fabrics and the ability to grow vining plants on the fence have given it some aesthetic uses. Chain-link fencing has grown in popularity to the point that the industry manufactures a wide range of easy-to-install parts, fixtures, and accessories. These ready-made components have made the construction of chain-link fencing easier than in years past.

The primary structural components of chain-link fence include heavy-duty posts, termed **end or corner posts**, that are set in concrete footings at the corners of installations and on either side of gates. Lighter-weight posts, termed **line posts**, are driven into the ground between the corner posts. The framework of the fence also includes a railing that runs across the tops of the posts called a **top rail**, and a **tension wire** which runs across the bottom of the fence. The top rail is installed using special caps, whereas the tension wire is installed using brackets with special hardware that bolts to the posts. To add strength and stability, a middle rail, or mid-rail, is installed on tall fences and at each panel adjacent to corners and gates of any fence over 4' tall.

After the framework is installed, the chain-link fabric is rolled out, stretched between posts, and fastened to the framework using metal ties. Gates included as part of the layout require that an opening be planned in the series of posts. A corner post is placed on either side of the gate and a hinged fabric panel hung on one of the posts.

Installing a chain-link fence

CAUTION

- Locate all utility lines prior to construction.
- Use caution when cutting materials.

Time: 30 minutes–1 hour per LF of fence.

Level: Challenging (42 steps). Lifting required. Two people may be needed to complete the project.

continued

Tools Needed:

1. Plan for fence installation.

2. Marking paint.

3. Marking pen.

4. 25' tape measure.

5. 100' roll of mason's twine.

6. Carpenter's level.

7. Clamshell or auger fencepost excavator.

8. Fence-post driver.

9. Round-nosed shovel.

10. Wheelbarrow.

11. Hacksaw.

12. Claw hammer.

13. Screwdriver.

14. Pliers.

15. Fencing tool.

16. Socket-wrench set.

17. Wood concrete float.

Materials Needed:

1. Chain-link fence components, including:

   ■ Corner and end posts (minimum of two, plus one extra for every 50 LF of fence, one for each corner, and two for each gate).

   ■ Line posts (one for every 10 LF of fence).

   ■ Top railing (one LF for every LF of fence).

- Mid-railing (40 additional LF of top railing).
- Line-post caps for top rail (one per line post).
- Corner-post caps for top rail (one per corner post).
- Rail holder fitting (two per mid-rail).
- Tension-wire connectors (2 per corner post).
- Tension wire (1 LF for every LF of fence).
- Chain link fabric (one LF for every LF of fence).
- Tension bars (two for every corner post plus two extra).
- Fabric ties (two per LF of fence).
- Premanufactured gate with hardware (one set per gate).

2. Angular 1" crushed stone (1/2 CF per post).

3. Concrete for corner and end posts (two CF per corner post).

Directions:

Corner-Post Installation:

1. Following your plan, layout the location of each proposed fence post. Using an auger or by hand, excavate an 8" diameter hole to 6" below maximum frost depth at each corner- and gate-post location. Holes should be no less than 3' deep for corner- and gate-posts. If your fence is longer than 50 LF, install a corner post midway, or every 50', whichever is less.

2. Mark posts using a permanent marker with a line at the correct burial depth (Figure 8-17).

3. Fill the holes with a stiff concrete mix and set the corner- and gate-posts in the hole. If the posts sink, remove them and wait 15 minutes. When the concrete has hardened enough, the post will sit in the footing without sinking. Twist the post down to the mark and check for plumb.

continued

4. Repeat for all corner and gate posts. Wait 48 hours for the concrete to completely set before continuing fence construction.

Line-Post Installation:

1. Between the tops and bottoms of corner posts, connect a string-line for aligning the fence. Mark the location of each line post. Typical spacing for panels is 10 feet, but spacing can be reduced to maintain even panel dimensions.

2. Using a fence-post driver, drive a line post at each location marked along the alignment (Figure 8-18). Check often for plumb and alignment.

Figure 8-17

The marking on the corner post indicates proper setting depth in the concrete footing. Check shortly after setting to ensure the post has not moved.

A less stable alternative to driving the line posts is to dig or auger holes and place the posts in the holes. Partially backfill the holes with gravel and adjust the height to match the stringline. Check for level in each direction and continue to backfill with soil or with concrete. If backfilling with soil, compact after every 6" of fill. Line posts can also be set in concrete, but doing so is only really necessary in poor soils or installations that require a high degree of stability, such as athletic fields or security installations.

Framework Installation:

1. Install the special caps on each corner, gate, and line post. Corner-post caps, which fit on top of the corner posts, have openings that accommodate the end of a railing, whereas the line-post caps slip on top of line posts and have an opening through which the railing runs.

Figure 8-18

Chain link fencing: line-post driving.

2. Top rails have a tapered end designed to join with another section. Begin installation by sliding the non-tapered end through the line-post caps and into one corner-post cap (Figure 8-19).

3. Place a second section of top railing through the next line-post cap and join it with the first section by sliding the non-tapered end over the tapered end of the first section of top rail.

4. Continue placing the top rail until a section passes over the corner post at the other end of the run.

5. Hold the top rail against the corner-post cap and mark where the end of the rail will meet the back of the opening in the cap. Using a hacksaw with a metal cutting blade, cut the top rail at the mark. Lift the corner-post cap and insert the cut end of the top rail in the opening. Lower the cap back onto the corner post.

continued

6. Continue installing top rail between all pairs of corner- and end-posts.

Tension-Wire Installation:

1. Install the tension-wire clamps around the base of each corner post.

2. Connect the tension wire to one clamp and stretch tightly between the two corner posts.

3. Attach it to the clamp at the base of the second corner post.

Figure 8-19

Chain-link fencing: installing top rail through special caps for line posts.

Mid-Rail Installation:

1. Identify a fencing panel where mid-rail is to be installed. On the post at one side of the panel, securely bolt the rail holder fitting half way up the post. At the same level on the second post, bolt another fitting that is loose enough to slide up and down.

2. Measure and cut a piece of top rail to fit between the first and second post of the panel. Be sure the measurement takes into account the fitting on each end.

3. Insert the cut rail into the first fitting. Slip the end of the loose fitting over the other end of the mid-rail. You may need to slide the fitting up or down to allow the rail to slip into the fitting.

4. After inserted, slide the rail and fitting up or down until the mid-rail is at the correct level. Securely bolt the second fitting to the second post.

5. Repeat this process for every panel that requires a mid-rail.

Fabric Application:

Fabric for the chain link fence will be stretched by hand for this installation. It is not uncommon for fabric that is stretched tightly by hand to stretch another 1'-2' when a professional uses a jack or a mechanical **stretcher**. Fabric can be stretched over straight runs and around the outside of curved installations. If the fence runs up a slope, set corner posts at the top and bottom of the slope and stretch separately for this section. Inside curves can only be stretched in short runs.

Install the fabric using the following steps:

1. Install by rolling the fabric out along the fencing run and leaning it against the posts to roughly determine the length of fabric needed (Figure 8-20). Pieces of fabric can be spliced together by overlapping loops of the fabric and placing a **stretcher bar** through this overlap (Figure 8-21).

2. At one end of the run, insert a stretcher bar through the first loop of the fence.

3. Fasten clamps around this stretcher bar and the corner post. Install one clamp for each foot of fence height (Figure 8-22).

4. Stretch the fabric as tightly as possible by hand along the entire run. Mark where the new end to the fabric should be after stretching.

5. Shorten the fabric to the desired length by laying it on the ground and doing the following:

   ■ Disconnect a fabric link at the new end by straightening the bend at the top and bottom of the link (Figure 8-23).

   ■ With a twisting motion, spin the disconnected loop out of the fabric to separate the fencing (Figure 8-24).

6. Insert a stretcher bar through the last loop of the fabric.

continued

Figure 8-20

Chain-link fencing: fabric layout.

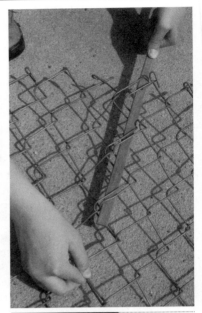

Figure 8-21

*Chain-link fencing: joining
sections using stretcher bar.*

7. Lean the fabric back up against the fencing framework.

8. Pull the end of the fabric by the stretcher bar until it reaches the corner post. Install clamps around the corner post and stretcher bar. Install one clamp for each 1' of fence height from the fabric.

9. Complete connections at the post (Figure 8-22).

10. Attach the fabric to the top rails, posts, and tension wires using aluminum fence ties (Figure 8-25). Work from the side of the fence opposite the fabric. Place the hook end of the tie over a strand of the fabric. Holding the hook in place, use one finger to bend the tie around the post or rail. Push the straight end of the tie back through the fabric and bend it around a strand of fabric.

Figure 8-22

Chain-link fencing: completed end connection showing top rail, end post cap, stretcher bar, and connector clamps.

Figure 8-23

Chain-link fencing: disconnecting top link so fabric can be shortened.

11. Repeat tie installation every 2' up and down the posts and every 4' along the top rail and tension wire.

Premanufactured Gates:

1. Verify the dimensions of the gate, including hardware, before setting the gate posts.

2. Install hardware and hang the gate.

continued

Figure 8-24

Chain-link fencing: separating links by twisting the disconnected link.

Figure 8-25

Chain-link fencing: connecting fabric to fence with ties.

Dry-Laid Fieldstone (Rubble) Free-standing Walls

Walls differ in many ways from fences but can still perform the function of creating and separating space. Because of their construction, walls require more structure to support their weight and resist wind. Free-standing walls are typically random-pattern installations, built either with or without mortar. As with fences, wall height will determine the primary function served by the wall. Short walls do an excellent job of defining space, and tall walls are an excellent screen and perform well in articulating space. Although fences up to six feet tall can be built by the homeowner, the techniques for constructing walls over three feet tall should be reserved to the experienced mason.

Fieldstone, or rubble, is the term for stone weathered by natural forces. When used in rustic, rural settings, the fieldstone wall is one of the most attractive landscape additions possible (Figure 8-26). Such walls are primarily used for boundary identification, with screening and enclosure being supplemental functions. Sound installation of free-standing walls requires establishing a stable base on which to place the wall. In addition to the base, careful selection of the material and staging of the pieces for arrangement is required. Obtaining the pattern and fit that the skilled craftsman can deliver takes patience and practice on the part of the homeowner.

Figure 8-26 | *Dry-laid stone wall.*

Installing a free-standing stone wall

CAUTION

- Locate all utility lines prior to construction.
- Walls over 36" high should be constructed by a contractor.

Time: 1–2 hours per LF of wall.

Level: Challenging (8 steps). Heavy lifting required. Artistic selection and placement of stone required.

continued

Tools Needed:

1. Plan for project.

2. Marking paint.

3. 25' tape measure.

4. Stone hammer.

5. Rubber mallet.

6. Round-nosed shovel.

7. Garden rake.

8. Wheelbarrow.

9. Torpedo level.

10. Hand tamper.

11. Optional: vibratory plate compactor.

Materials Needed:

1. Wall stone. To estimate the amount of stone needed, calculate the surface area of the wall. A typical installation for a wall 1' thick will require 1 ton of stone for every 30 SF of wall. Thicker walls will require more stone. Contact your stone supplier to confirm the quantity required.

2. Angular 1" crushed stone base. To calculate the CF of stone required, multiply the wall length by the wall width plus 1'. Divide your answer by 2.

Directions:

Base Preparation:

1. Paint the location of the wall alignment on the ground. Add 6" to the front and back of the alignment to allow for the width of the base trench.

2. Excavate a base trench 6" deep and 12" wider that the wall width along the entire wall alignment.

3. Fill with angular 1" crushed-stone base material and compact it using the hand tamper or vibratory plate compactor. Adjust the surface until it is close to level.

Stone Placement:

1. Lay out several stones and sort and place by hand. Stone placement requires sorting and fitting of materials into available openings in the wall. All sizes of stone, including small pieces, may be used in the construction.

2. Install course one by placing larger, flat stones on the base material. Course one should be at least twice as wide as the top course (Figure 8-27).

Figure 8-27 *Dry-laid stone wall: stone placement.*

continued

3. Continue to add courses. Intermingle stones of varying sizes as the wall height increases. Stones should occasionally be placed perpendicular to the face of the wall to stabilize the wall (Figure 8-28). Additional stability can be achieved by placing stones in a tapered or pyramidal cross-section, with subsequent courses narrowing slightly as wall height increases. Small pieces of stone can be used to shim and level larger stones, and to fill small voids when larger stones do not fit perfectly.

4. Periodically step back and view the wall to judge the esthetics of your work. In addition to varying stone sizes, mixing the colors and textures of the available stone will add to the beauty of the wall. Reset sections that lack interest.

5. Cap the wall using wide, flat stones if available.

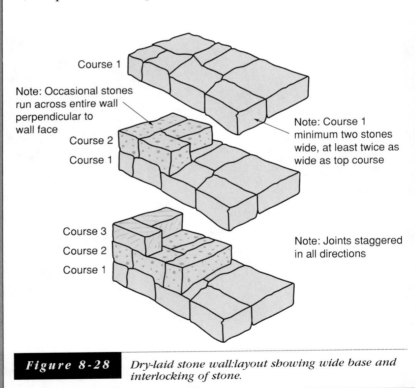

Course 1

Note: Occasional stones run across entire wall perpendicular to wall face

Course 2

Course 1

Note: Course 1 minimum two stones wide, at least twice as wide as top course

Course 3

Course 2

Course 1

Note: Joints staggered in all directions

Figure 8-28   *Dry-laid stone wall:layout showing wide base and interlocking of stone.*

Chapter 9

WATER, EDGING, LIGHTING, AND OTHER SITE AMENITIES

The foundation of your landscape design is the arrangement of major landscape elements, but the flair of design relies on choice of placement of the items that embellish the site. Items such as walks, patios and decks, fences, and walls create a framework for function and aesthetics on a site, but details such as lighting, water features, and edgings add the accents that distinguish your landscape. When creatively blended with plantings, each of these elements contributes to the overall quality of the site.

The choices for your site accents, or amenities, can be overwhelming. Any item has the potential to add to the design scheme of a residence. And for each item that can be used as an amenity, there are typically many choices of materials, colors, forms, and textures. Selecting the amenities that will perfectly enhance the look of your site will require many hours of searching and evaluation. As with all other aspects of landscape construction, the time and effort you invest in planning and preparation will profoundly impact the attractiveness of your site.

PLANNING FOR SITE AMENITIES

Whether installing your amenities requires only setup or extensive utility lines and foundations, the key to success is anticipating how the amenity will function during all seasons, all conditions, and at various times of day. Without planning, your dreams of an amenity-enhanced site can be dashed because the size, configuration, or appearance are not appropriate to the site and the intended use.

Locating Amenities

The placement of site amenities should accent the landscape and, if possible, address functional needs. Items such as stairs and lighting may be located where they improve access to the landscape, whereas pools and edging may be placed solely for aesthetic reasons. Consider the requirements of your amenity when selecting a site. Pools will require a flat area, whereas water cascades work best with a moderate slope. Steps should be placed where they help you effectively traverse slopes, and seating will require a level location. Lighting and edging can be placed in a variety of slope conditions and situations.

Legal issues may also enter into the planning and placement of a site accent. Check with local officials for any rules regarding the installation of pools and water features. Any site amenity that requires the use of utilities will fall under the jurisdiction of building codes and possibly zoning and other municipal regulations, in order to protect public health and safety.

Utility Hookups

Several amenities require access to utilities to be successful. If you plan an installation that requires electricity, such as a pump or lighting, be certain you have eliminated the risk of shock. If your project requires the installation of an electrical circuit, obtain assistance from a professional. Circuits for all exterior work should be **GFCI** (ground fault circuit interrupt) protected. GFCI circuits are designed to shut off electrical power to the circuit when voltage fluctuations from short circuits are sensed.

CAUTION

- When working with electricity in any form, high or low voltage, there is the danger of electrical shock. If you are not familiar with electricity, hire a licensed electrician to do this work. Many communities and locations require that all wiring by done by a licensed electrician, even if the homeowner is knowledgeable about electricity.

- Installing water features will require access to a water supply. Some low-volume features may be initially filled using a hose from a nearby hydrant or faucet, with the water level occasionally replenished from the same source. Large water installations, those in warm climates, and features that need the water periodically changed may require that a water-supply line be run to the installation. If you will need a water line extended to your feature, have the installation completed by a licensed plumber prior to beginning your work.

In climates that face cold temperatures, you will also need to consider protecting your water supply from freezing. Burial to frost depth will be required, as will installation of freeze-proof valves and insulating components that are above the frost line. Plan how you will drain a feature before its installation, either via a pump or by installing a drain. Water removed from the feature will have to be disposed of in a manner that does not damage the site or flood the neighborhood.

CAUTION

■ Any water feature that is directly connected to a potable water line must have a **backflow prevention valve** installed between the feature and the water source to prevent contamination of the water supply. Check with local building officials regarding regulations on the placement of such a valve.

■ If you fill a pool or fountain with a hose, never leave the hose end under the surface of the water; this could result in the siphoning of the water back into the municipal water supply, which could contaminate it.

Foundations for Amenities

Many amenities can be set directly on finish grade with minimal preparation of a base. In some cases, it may be sufficient to install a granular material as support for your feature. In others, pouring a concrete slab or setting precast concrete blocks may be required, to provide the necessary support and stability needed for seating and other permanent amenities.

WATER FEATURES

Whether your design relies on reflections in a pool or the sound and motion of a fountain, water features can add interesting dynamics to the landscape. New construction materials have made the introduction of water into the landscape a feasible undertaking. Traditionally, landscapes have relied on construction using concrete-lined pools with underground plumbing. Although this construction technique is still used, the introduction of **flexible-pool liners**, small pumps, and plastic tubing has made introducing water features into the residential domain much easier.

Although today's technology has allowed for water features to be easily implemented in landscapes, they nevertheless require high maintenance. In all climates, you must make a commitment to regulate water quality, and in some climates, pools will require periodic draining if the feature is to stay attractive and functional.

Pools

Pools are water reservoirs with little or no moving water. Pools can be constructed using either flexible vinyl liners, which allow you to determine size and shape, or with rigid liners that have a set form and size. To hide the liner, the edge of the pool is covered with **coping**—that is, flat pieces of stone or a similar material. When considering both the initial installation and the long-term maintenance of a pool, you should spend ample time planning the installation. Using either your own creativity or the assistance of a designer you can create a water feature that will compliment any design. You can enhance the appearance of your pool with landscaping around the edges or by submersing aquatic plants in the pool. You can further enhance it by adding boulders at locations around the edge or on a **shelf** in the pool. Placing the pool where it catches a water **cascade** or adding a **fountain** will add motion to your creation.

Flexible-liner installations will be suitable for anyone who plans to significantly vary the shape and depths of their pool, but such liners require more effort in fitting the liner to shape of the basin. Rigid liners are available in many shapes that readily fit a prepared excavation, but the designer cannot change the dimensions of the liner that is purchased. Whichever liner type you plan to use, the entire perimeter must be perfectly level. Variations in grade around the perimeter will cause water to drain out of the pool at the low points and expose the liner on the high side. Both liners types are reasonably durable and, with proper installation, should last for many years.

Installing a flexible-liner pool

Note: The shape, size, and depth of the pool you construct can be altered from the plan shown in this project. If you do alter the design, however, be certain that the liner you purchase is large enough for the pool.

CAUTION

- Verify the location of utility lines before construction.
- Follow the manufacturer's instructions when using equipment.
- Use caution when cutting and installing materials.

Time: 6 hours to 2 days, based on the complexity of the pool.

Level: Challenging (13 steps). Digging and heavy lifting required.

Tools Needed:

1. Plan for pool installation.

2. Marking paint.

3. Round-nosed shovel.

4. Square-nosed shovel.

5. Carpenter's level.

6. Straight 2×4 the same length as your pool width.

7. Brick hammer.

8. Rubber mallet.

9. Sponge.

10. Garden hose(s).

11. Wheelbarrow.

12. Location for disposing of excavated soil.

continued

Materials Needed:

1. Flexible liner. To determine the proper size of the liner use the following formulas.

   ■ Liner length in feet: L + 2D + 4 where L = pool length at the rim and D = pool depth.

   ■ Liner width in feet: W + 2D + 4 where W = pool width at the rim and D = pool depth.

   ■ For irregularly shaped pools, measure the length, width, and depth using maximum possible dimensions.

2. 3 CF sand (more for large pools) to even the interior of the pool excavation.

3. Enough coping to circle the entire perimeter of the pool. Pieces of flat, square stone or precast concrete at least 2" thick and 12" wide and long make the best coping material.

4. Water to fill the pool. This may require running a hose from a remote location or installing a permanent water-supply line.

Directions:

1. Review the location of your pool on the plan.

2. Using the paint, mark the location of the pool basin on the ground (Figure 9-1, Step A).

3. Excavate the pool basin to proper depth. Steep sidewalls work best for flexible-liner installations. If different depth levels within the pool are desired, adjust the excavation where these "shelves" should occur (Figures 9-1, Step B, and Figure 9-2).

4. Excavate a ledge for coping around the entire perimeter of the basin. The edge should be the same thickness and slightly narrower than the material you have selected for coping. It is critical to the success of the installation that the entire perimeter be **level**. Verify the ledge is level along the entire perimeter (Figure 9-1, Step C). Adjust grade if necessary. Level across the pool can be checked by placing a

Step A. Mark pool basin location.

Step E. Center liner over pool basin and push into basin. Smooth and adjust liner location.

Step B. Excavate pool basin to proper depth with proper side angle. Shelves of varying levels can be created by adjusting the depth of the excavation.

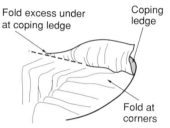

Fold excess under at coping ledge

Coping ledge

Fold at corners

Step F. Fold excess liner under at back edge of coping ledge.

Ledge to match coping dimensions

Ledge must be level along the entire perimeter

Step C. Excavate ledge for coping.

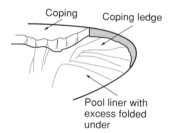

Coping

Coping ledge

Pool liner with excess folded under

Step G. Install coping.

Step H. Clean and fill pool.

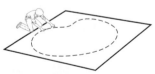

Step D. Lay out liner and check for proper dimensions.

Figure 9-1 *Pool excavation and liner installation. Pools can be built in many shapes.*

continued

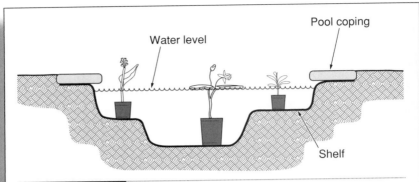

Water level

Pool coping

Shelf

Figure 9-2 *Varying pool depth to create shelves.*

straight 2 × 4 on edge across the pool and setting a carpenter's level on top.

5. Irregularities in bottom of basin walls can be corrected by filling low areas with wet sand.

6. Lay out the liner and check for proper dimensions (Figure 9-1, Step D).

7. Center liner over basin opening and push liner into basin (Figure 9-1, Step E). Stepping into the pool may assist in positioning the liner (Figure 9-3); if you do, however, take care that you don't tear the liner. Adding a few inches of water to the pool at this point will also assist in forming the liner to the bottom of the excavation.

Figure 9-3 *Installation of liner in a flexible-liner pool.*

8. Adjust the liner to cover entire basin and fit it snugly over any shelves and the coping ledge. Smooth as many wrinkles from the basin as possible.

9. Verify again that the coping edge is level around the entire perimeter. Adjust if necessary.

10. Fold any excess liner back under the pool at the coping ledge. This will provide additional liner in the event adjustments are necessary (Figure 9-1, Step F).

11. Place coping on the coping ledge. The coping should overhang the edge 2–3". This overhang will hide the corner of the pool and the top part of the liner (Figure 9-1, Step G, and Figure 9-4). Coping may need to be shaped using the stone hammer. Use a rubber mallet to fit pieces.

12. Clean the inside of the pool with water and a sponge (Figure 9-1, Step H).

13. Fill the pool to within 2" of the bottom of the coping.

Figure 9-4   *Edging a flexible-liner pool. Coping hangs over the water to hide the liner edge.*

Installing a rigid-liner pool

CAUTION

■ Verify the location of utility lines before beginning construction.

Time: 6 hours to 2 days; varies based on the complexity of the pool.

Level: Challenging (10 steps). Digging and heavy lifting required.

Tools Needed:

1. Plan for pool installation.

2. Marking paint.

3. Round-nosed shovel.

4. Square-nosed shovel.

5. Carpenter's level.

6. Stone hammer.

7. Rubber mallet.

8. Straight 2 × 4 the same length as your pool width.

9. Sponge.

10. Garden hose(s).

11. Water.

12. Wheelbarrow.

13. Location for disposing of excavated soil.

Materials Needed:

1. Rigid pool liner.

2. 3 CF of sand (more for large pools) to even the interior of the pool excavation.

3. Enough coping to circle the entire perimeter of the pool. Pieces of square, flat material at least 2" thick and 12" wide and long make the best coping material.

4. Water to fill the pool. This may require running a hose from a remote location or installing a permanent water line.

Directions:

1. Review the location of the pool on your plan.

2. Using the paint, mark the location for the pool basin. Trace a paper outline of the pool if necessary to get a precise location of the basic shape and size (Figure 9-5, Step A).

3. Excavate the pool basin to the proper depth. Sidewall angles should match the angles on the pool (Figure 9-5, Step B).

4. Excavate a ledge for coping around the entire perimeter of the basin. The edge should be the same thickness and slightly narrower than the material you have selected for coping. It is critical to the success of the installation that the entire perimeter be level. Verify that the ledge is level along the entire perimeter. To check the level across the pool, set a straight 2 × 4 on edge and place a carpenter's level on top. Adjust grade if necessary (Figure 9-5, Step C).

5. Irregularities in the bottom or sides of basin walls can be corrected by filling holes with wet sand.

6. Lower the liner into basin to check for fit. The pool liner should fit snugly into the basin excavation and the top of the liner should match the coping ledge. If adjustments are required, remove the liner and make them (Figure 9-5, Step D).

7. Verify again that the coping edge is level around the entire perimeter. Adjust if necessary.

8. Place the coping on the coping ledge. Let the coping hang over the edge 2–3". This will hide the corner of the pool

continued

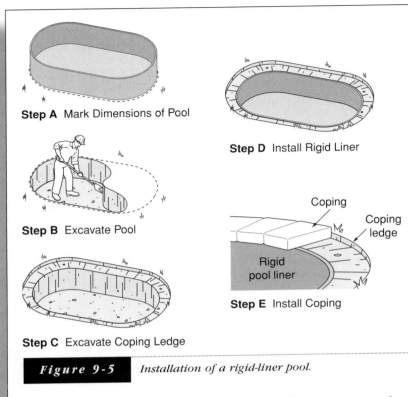

Step A Mark Dimensions of Pool

Step B Excavate Pool

Step C Excavate Coping Ledge

Step D Install Rigid Liner

Coping

Coping ledge

Rigid pool liner

Step E Install Coping

Figure 9-5 *Installation of a rigid-liner pool.*

and top of the liner (Figure 9-5, Step E). If necessary, use the stone hammer to shape the stone.

9. Clean the inside the pool with water and a sponge.

10. Fill the pool to within 2" of the bottom of the coping.

Fountains

Fountains are pumps with special **orifices** that project water upward out of a reservoir. Depending on the size of the pump and the type of orifice, water may project a few feet into the air or simply bubble to the surface. Different types of fountains can be combined and lighting added for special effects. Fountains are typically set just below the surface of the water in a pool to hide their mechanical components. Electrical connections run from the fountain to the edge of the pool, where it can be threaded through the coping and into a source of electricity.

Installing a pool fountain

CAUTION

■ Use caution when working with electricity. Only equipment approved for water locations should be used in a pool.

■ All circuits should be turned off when installing and checking fixtures.

■ Circuits must be ground fault (GFCI) circuits.

■ Do not operate a submersible fountain unless it is under water.

Time: 2–4 hours.

Level: Easy (8 steps).

Note: You must have a water feature in which to place your fountain.

Tools Needed:

1. Electrical cord.

Materials Needed:

1. **Submersible fountain**.

2. Desired orifice (choices will include bubblers, sprayers, and streamers). Some fountains have a nozzle built into the fountain.

3. Stone or block of various sizes.

4. GFCI electrical source for fountain operation.

continued

Directions:

1. Review the manufacturer's instructions for the correct depth at which the fountain should be set.

2. Drain the water from the pool to a depth that allows you to wade into the pool to work; take care not to tear the liner with your shoes or tools.

3. Stack stone or block in a stable arrangement on the bottom of the pool to a height that will position the fountain at the correct depth. The fountain may also be set on a shelf that was excavated during pool installation.

4. Set the fountain on the stack or shelf (Figure 9-6).

5. Route the power cord for the fountain to the edge of the pool, between or under pieces of coping, and out to a GFCI outlet or to power cord connected to a GFCI outlet. The cord should be routed so it is hidden from view. No electrical connections should be made in wet areas.

6. Refill the pool to the maximum level.

7. Plug in and test the fountain.

8. Unplug the electrical cord and correct any problems with the fountain or orifice.

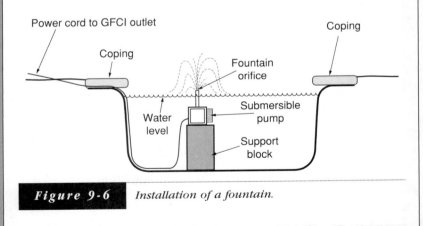

Figure 9-6   *Installation of a fountain.*

Cascades

A cascade is water that gently flows from a higher elevation to a lower elevation, providing the sight and sounds of running water. A shallow channel, lined with a strip of flexible pool liner topped with stone or gravel, confines the water as it flows between elevations. To recirculate, a pool or reservoir at the bottom of the cascade catches water and a pump returns it to the top of the cascade (Figure 9-7). Cascades work best on shallow slopes blended into the existing landscape with boulders and plantings. You can greatly enhance the appearance of a cascade by adding low-voltage waterproof lighting to highlight the water movement at key locations along the cascade.

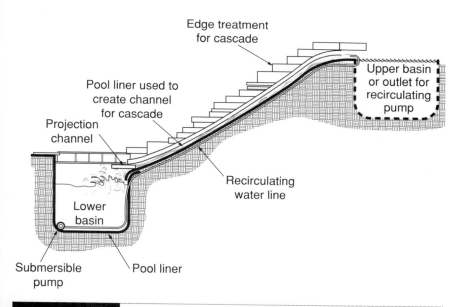

Figure 9-7   *Cascade cross-section showing water source and basin.*

Installing a stone-lined cascade

Time: 4 hours to 2 days for a cascade approximately 10 feet long.

Level: Moderate (8 steps). Digging and heavy lifting required.

Note: For the cascade to function properly, you must provide a water supply and a water-collection point. A pool may serve as the collection point, with a submersible recirculating pump providing the continuous supply of water.

CAUTION

- Verify the locations of all utility lines before beginning construction.
- Use caution when working with electricity.
- All circuits should be turned off when installing and checking fixtures.
- Circuits must be ground-fault-interrupt (GFCI) circuits.
- Do not operate a submersible fountain unless it is under water.

Tools Needed:

1. Plan for project.
2. Marking paint.
3. 25' tape measure.
4. Round-nosed shovel.
5. Wheelbarrow.
6. Carpenter's level.
7. Utility knife.
8. Caulking gun.
9. Screwdriver.

Materials Needed:

1. Flexible pool-liner material. To calculate the amount required measure the length of the channel. You will need a strip of liner that is 4' wide and the length of the channel. You can overlap multiple strips to create the liner, their length depending on the length of the cascade and the number of changes in direction.

2. Stone for lining and edging the cascade. To determine the amount of stone required, purchase approximately six 12" × 12" pieces for each LF of channel. Smaller stone and washed gravel will also aid in decorating the channel.

3. 1 tube of clear silicon caulking.

4. 1 submersible pump and enough tubing to run from the pump to the top of the cascade plus 10'.

Directions:

1. Review the location for the cascade on the plan.

2. Using the paint, mark the location of the cascade channel on the ground. The channel may follow any alignment desired, as long as the proper cross-section and downward slope are maintained.

3. Excavate an 8" deep by 14" wide trench along the entire route of the channel, maintaining a consistent slope from top to bottom. Angle the edges of the channel upwards to confine the water within the channel.

4. Waterproof the channel by cutting a linear strip of flexible liner material and placing it in the excavated channel. If splices need to be made or corners must be turned, overlap two pieces of liner approximately 1 foot, placing the upper section on top of the lower section. Where the cascade reaches the lower collection pool, let the liner drape over the pool edge. Lay a bead of silicon caulking where the liners overlap to reduce leaking and wicking (water being drawn up between surfaces).

continued

5. Place stone lining in the channel, beginning at the bottom. The bottom piece should overhang the collection point by 2–3". Stones should be placed firmly on the bottom of the channel, level side to side, and with a slight tilt towards the downhill side of the channel.

6. Place the submersible pump in the collection pool. Route the recirculating water line along the side of the cascade to the top.

7. Fill the collection reservoir and test the pump to verify that the cascade is working properly. If the volume is inadequate, install larger or multiple pumps and recirculating lines to increase the water flow.

8. Treat the edges by stacking stone pieces 2 to 3 high along the edge of the channel. Let the edge stones overlap the channel bottom stones. Careful placement of plant material can hide wiring, tubing, and other rough edges of the cascade installation (Figure 9-8). Do not crush or crimp the recirculation tubing when placing edge treatment.

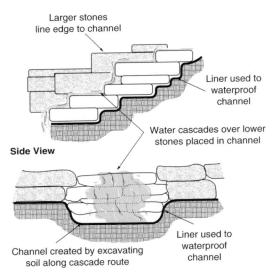

Larger stones line edge to channel

Liner used to waterproof channel

Water cascades over lower stones placed in channel

Side View

Channel created by excavating soil along cascade route

Liner used to waterproof channel

Front View

Figure 9-8 *Cascade edge treatment front and side view.*

EDGING

Edging is one of the most versatile, yet subtle, materials in the landscape. Edging can be used to define areas, hold pavement in place, and separate turf from planting beds. Edging materials range from plastic to stone, and installation for most is as simple as excavating a shallow trench and placing the material. Most edging is purchased as either individual units placed one at a time or long strips staked into the ground. Although installing individual-unit edging requires more time, it results in a more aesthetically appealing feature, whereas strip edgings provide the necessary separation in a short amount of time. Almost all edging types can be shaped to match any line required in the landscape.

Natural Edging

A simple edge that conforms to any shape and requires little special preparation is the natural edge. Created by cutting a vertical trench with a shovel, natural edges isolate the planting bed from the lawn without a manufactured separator. However, as turf encroaches into the planting bed natural edging will require periodic maintenance to keep them defined.

Natural edging

Time: 1 hour for every 20' of edging required

Level: Easy (4 steps). Digging required.

Tools Needed:

1. Plan for the project.

2. Marking paint.

3. Trenching shovel.

4. Wheelbarrow.

continued

Directions:

1. Identify the location of the edging on the plan.

2. Using the paint, mark the entire alignment of the edging on the ground (Figure 9-9).

3. Along the entire alignment, excavate a shallow trench approximately 4–6" deep. The lawn side edge of the trench should be excavated with a vertical edge, whereas the planting-bed side of the trench should be excavated at a 45°angle (Figure 9-10).

4. Fill the planting bed with mulch to within 1" of the top of the trench.

Figure 9-9

Layout of edging using marking paint.

Figure 9-10

Proper trench edging showing vertical excavation on lawn side.

Plastic Edging

One of the most commonly used edging materials is black plastic manufactured in rolls and long strips. This edging is flat and flexible, so that it
can be bent to very tight radii, is held in place with metal stakes, and
has only the rounded bead on top visible above the ground after installation. Offsetting its ease of installation is plastic's reputation as unattractive. You may also discover additional shortcomings of plastic
edging when your mowing equipment damages the material and frost
pushes the edging up.

Installing plastic edging

CAUTION

Use caution when cutting edging.

Time: 2 hours for every 20 feet of edging required.

Level: Easy (11 steps). Digging required.

Tools Needed:

1. Plan for the project.

2. Marking paint.

3. 25' tape measure.

4. Trenching shovel.

5. Wheelbarrow.

6. Claw hammer.

7. Tin snips.

8. Hacksaw.

continued

Materials Needed:

1. Coils or strips of edging. To calculate the amount of edging needed, measure the length of the edging on the plan, and order 10 percent more. If using coils, straighten the edging by unrolling the coils and weighting the ends.

2. Edging stakes, 1 for every 2 LF of edger.

3. Joining tubes.

4. 2 8d galvanized box nails for every 10 LF of edger.

Directions:

1. Identify the location of the edging on the plan.

2. Using the paint, mark the entire alignment of the edging.

3. Along the entire alignment, excavate a shallow trench approximately 1" deeper than the width of the edging. Excavate the lawn side of the trench with a vertical edge and the planting-bed side at a 45° angle.

4. Place a length of the edging flat against the vertical side of the trench. If the edging has a fold of plastic, or a V-shaped channel, place the fold or channel toward the planting bed side of trench.

Figure 9-11

Staking plastic edging. Edging must be held flat against edge of trench.

5. Hold the top of the edging, usually the rounded bead, at the top of trench.

6. Place an edging stake in the V channel at the bottom of the edging and use a hammer to drive the stake through the edging into the subgrade. Verify that the top of the edging is still at the top of the trench (Figure 9-11).

7. Continue driving edging stakes every 2 feet along the length of the edging.

8. Bend the plastic edging around any corners. To prevent the edging from buckling when turning tight corners, cut 2" slits up from the bottom every 6".

9. Join pieces of edging by cutting 6" of the rounded bead off the top of one end of an edging piece. Insert a joining tube halfway into the rounded bead of that piece of edging. Overlap the flat sections of the first and second pieces of edging and slide them together (Figure 9-12). Slide the

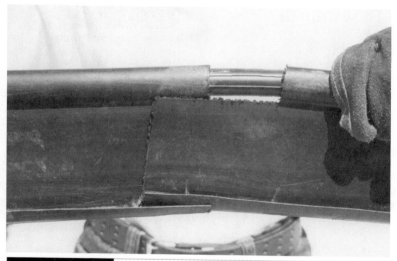

Figure 9-12 *Joining plastic edging using the tubing supplied with the edging kit. Cutting the round portion off of the last 4" of one piece of edging improves the fit.*

continued

remaining half of the joining tube into the rounded bead on the other piece of edging. Drive an 8d nail through the flat portion of the overlapped pieces.

10. If smaller pieces of edging are required, measure and cut them with a hacksaw or tin snips (Figure 9-13).

11. Backfill and compact the soil along the planting bed side of the edging.

Figure 9-13

Cutting plastic edging. Tin snips can also be used.

Bender-Board Edging

Bender board provides a decay-resistant wood edging that is flexible enough to be installed around corners. Constructed of overlapping layers of thin wood that slide past each other when bent, bender board provides an effective and durable method for edging plantings and paving.

Stone Edging

Wall stone or flagstone can be arranged to provide you with an appealing and highly visible edging for planting beds. Using stone will require site preparation somewhat different than needed for other edging materials and will be costlier than most but is relatively more attractive and lower in maintenance.

Installing bender-board edging

Time: 2 hours for every 20' of edging required.

Level: Easy (10 steps). Digging required.

Tools Needed:

1. Plan for the project.

2. Marking paint.

3. 25' tape measure.

4. Trenching shovel.

5. Wheelbarrow.

6. Two-pound sledge.

7. Claw hammer.

8. Carpenter's saw.

Materials Needed:

1. Strips of edging. To calculate the amount of edging needed, measure the length of the edging on the plan. Order 10 percent more.

2. 10 16d nails for each 10' of edger.

3. Treated 2" × 2" × 12" long wood stakes.

Directions:

1. Identify the location of the edging on the plan.

2. Using the paint, mark the entire alignment of the edging on the ground.

3. Along the entire alignment, excavate a shallow trench approximately 1" deeper than the vertical dimension of the edging.

continued

Excavate the lawn side of the trench with a vertical edge the planting bed side of the trench at a 45° degree angle.

4. Place a length of the edging flat against the vertical side of the trench. Bend the edging to match the shape of the vertical side.

5. Hold the top of the edging at the top of trench.

6. Place a stake next to the edging and use the sledge to drive the stake into the subgrade. The top of the stake should be driven 1" below the top of the edging. Verify that the top of the edging is still at the top of the trench (Figure 9-14).

7. Install a nail through the stake into the edging to hold the edging in position.

8. Continue driving edging stakes every 4' along the length of the edging.

9. If smaller pieces of edging are required, measure and cut them with a carpenter's saw. Be certain to cut only after the edging has been bent to the desired curvature, because bending will alter the length required.

10. Backfill and compact the soil along the planting-bed side of the edging.

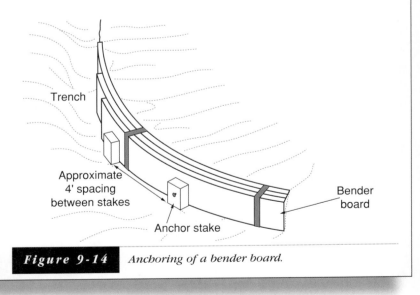

Trench

Approximate
4' spacing
between stakes

Anchor stake

Bender
board

Figure 9-14 *Anchoring of a bender board.*

Installing flat-stone edging

CAUTION

Use caution when shaping stone and cutting the weed barrier.

Time: 2–3 hours for every 20' of edging required.

Level: Easy (9 steps). Digging required.

Tools Needed:

1. Plan for project.

2. Marking paint.

3. Square-nosed shovel.

4. Wheelbarrow.

5. Brick hammer.

6. Scissors or utility knife.

Materials Needed:

1. Pieces of flat stone approximately 2" thick and 12" in the longest horizontal dimension. To calculate the amount of edging needed, measure the length of the edging on the plan. Order 10 percent more.

2. Optional: weed barrier cut into 12" wide strips. Weed barrier is fabric woven from plastic or heavy exterior fibers designed to prevent weeds from growing through the barrier.

Directions:

1. Identify the location of the edging on the plan.

2. Using the paint, mark the entire alignment of the edging on the ground.

continued

Figure 9-15 *Edging with stone placed on weed barrier.*

3. Along the entire alignment, excavate a shallow trench 14" wide, approximately 2" deep on the lawn side, and sloped up at a 30° angle toward the planting bed.

4. Smooth the surface of the trench.

5. If weed protection is desired, place the weed barrier along the length of the sloped trench.

6. Set a piece of stone in the trench with the straightest edge along the lawn side. Shape with stone hammer if necessary (Figure 9-15).

7. Place a second piece of stone adjacent to the first. Select a piece that fits the edge shape as tightly as possible, shaping if necessary. The edge forming on the planting-bed side can be irregular (Figure 9-16).

8. Continue placing stones until the entire bed has been edged.

9. On the planting-bed side, add mulch to the top of the edging covering the weed barrier.

Figure 9-16

Edging with flagstone placed on a fabric weed barrier. The straight edge of the stone is placed on the lawn side, the irregular edge on the bed side.

Vertical Brick Edging

Vertically oriented bricks can be used for edging paved surfaces or formal planting areas. Stood on end, on edge, or laid flat, the bricks introduce earthy textures and colors into the design that can be used to match a paving or building material. Brick edging is more expensive and labor intensive than other types of edging, but the aesthetic appeal may be worth the investment.

Installing vertical brick edging

Time: 2–3 hours for every 20 feet of edging required.

Level: Easy (6 steps). Digging required.

Tools Needed:

1. Plan of the project.

2. Marking paint.

3. 25' tape measure.

4. Trenching shovel.

5. Square-nosed shovel.

6. Wheelbarrow.

7. Hand trowel.

Materials Needed:

1. Enough bricks to complete the edging. To calculate the
   number of bricks, needed, measure the length of edging to
   be installed. Multiply that number by 6 to obtain the
   approximate number of bricks required (six bricks, set on
   edge, per linear foot).

2. 1 CF of sand for every 10 LF of edging.

Directions:

1. Using the paint, mark the alignment for the edging on the
   ground.

2. Excavate an 8" deep by 6" wide trench along the entire
   alignment of the edging with the trenching shovel. Shape
   the trench using the square-nosed shovel. The trench must
   form a vertical edge along both sides of the trench. The
   bottom of the trench should be as level as possible.

3. Place 2" of sand along the bottom of the entire trench.

4. Place bricks vertically in the trench with the shortest dimension along the outside edge. Each brick should set flush against the previous brick (Figure 9-17).

5. Add or remove sand with the hand trowel to adjust the height of the bricks so that each is flush with adjacent bricks.

6. Immediately backfill and compact the soil along the edging.

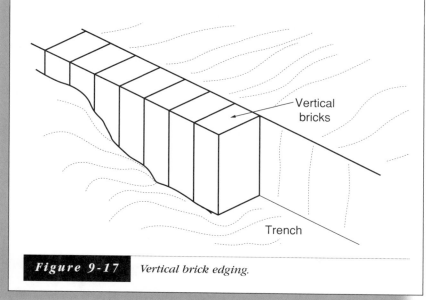

Vertical bricks

Trench

Figure 9-17 *Vertical brick edging.*

DC (DIRECT CURRENT) LIGHTING

Exterior low-voltage electrical systems are used primarily for powering your decorative lighting. Spotlights, stair lights, low-level walk lights, and uplights or downlights in plant material are examples of applications for direct-current lighting. **DC electrical** systems require an AC outlet to provide power. Typical AC sources are 110 volt outlets located in garages, basements, or exterior locations. You will need to locate an outlet in a weatherproof location where a transformer can be placed. It is necessary to select an outlet that is a ground-fault-circuit-interrupt (GFCI) circuit.

Most transformers are designed so that you can mount them on a post or wall with a grounded plug for an outlet. The **transformer** converts the AC current to DC current, so that it is usable by the low-voltage system. A low-voltage electrical cable plugs into the transformer and must be fed to the locations where lights are located. If the transformer is inside, the cable will need to be fed through an opening in a wall.

Some DC lighting systems have controllers that turn on lights at dusk or allow the user to control the times at which the systems turns on and off. These controllers may be built into the transformer or installed as separate units placed along the cable. Either method requires that the controllers be located where the owner—and sunlight, in the case of **photocells**—can gain access to the equipment. This may require transformers with built-in controllers to be located outside a structure and the cord that connects them to the AC outlet fed through the wall to the outlet.

Installation of basic DC lighting systems varies slightly from manufacturer to manufacturer, but the steps required for system assembly are similar.

Installing a direct-current (DC) lighting system

CAUTION

- Locate all utility lines before beginning construction.

- Use caution when working with electricity. Verify that electrical circuits have been turned off before working on lighting system.

- Exterior lighting systems should be connected only to groundfault-circuit-interrupt (GFCI) circuits.

■ Follow the manufacturer's instructions for all lighting installations.

Time: 2 hours for a five-light system. Transformer installation may require additional time.

Level: Moderate (15 steps).

Tools Needed:

1. Plan for the lighting.

2. Wire cutters.

3. Standard and phillips head screwdrivers.

4. Drill (cordless operation or electric).

5. 1" and 1.5" wood spade drill bits.

6. Claw hammer.

Materials Needed:

1. Caulking.

2. Connection to a GFCI 110-volt AC power source.

3. Pre-manufactured DC lighting kit containing the following items:

■ Transformer that can be plugged into a GFCI 110-volt AC-current outlet.

■ Cable.

■ Light units.

■ The kit may also have various controllers (timers, photocells, and so on) and additional light fixtures.

Directions:

Steps for installing DC lighting (Note: Instructions may vary slightly from manufacturer to manufacturer. Adjust the following

continued

steps to conform to the instructions supplied with your lighting kit.):

1. Place the light fixtures in the desired locations.

2. Locate your AC power source. Power source must be a 110V ground-fault-protected (GFCI) duplex outlet.

3. Mount the transformer near the selected outlet. If the AC power source is inside a house or garage, you may need to drill a hole through a wall and feed either the lighting cable or the power cord that connects to the AC outlet through the hole to reach the transformer. Before drilling any hole, verify that there are no structural, electrical, or other critical building components that you may inadvertently drill through. When finished, caulk or plug the hole with insulation or steel wool, which will prevent insects, rodents, and water from entering the cable hole. Fasten the cable to a nearby stud with a wire staple, taking care that the staple points are outside the insulation on the cable, so that they won't cause a short by coming into contact with the wire that carries the electricity.

4. Connect the controller to the transformer if it is a separate unit. Both should be placed where they will be easily accessible. If you have a controller with a photocell that turns lights on after dark and off during daylight), you will need to mount the photocell in an exterior location.

5. Lay the cable to the location of each fixture and connect fixtures as directed in the following steps.

6. Insert bulb into lamp base socket (Figure 9-18).

7. Attach the lens to the lamp base. Lenses will snap (bayonet mount) or twist (screw mount) into the base (Figure 9-19).

8. Run a loop of cable through the mounting-stem bracket (Figure 9-20).

9. Connect the lamp base to the cable by pressing the cable onto the metal prongs projecting from the lamp base. These prongs puncture the cable and make contact with each

Figure 9-18

Inserting a bulb into a lamp-base socket.

Figure 9-19

Attaching a lens to a lamp base.

conductor in the wire. One metal prong must make contact with the wire inside each side of cable. Use care not to bend the prongs (Figure 9-21).

10. Attach a threaded cap that will hold the cable in place (Figure 9-22).

11. Slide the mounting stem bracket into place over the threaded cap (Figure 9-23).

12. Insert the mounting stem into the mounting step bracket (Figure 9-24).

13. Gently push the fixture into the ground (Figure 9-25).

14. Plug the cable into the transformer, and then plug the transformer into the GFCI outlet.

continued

Connecting a cable to a lamp base. A metal prong must penetrate each side of the cable. Use caution to avoid bending the metal prongs.

Running a loop of cable through a mounting stem.

15. Test the system.

Ground-level lights may have the stem and fixture in a single unit, allowing you to connect the light and insert the mounting stem into the ground in a single operation. Mounted lights, however, may require that the fixture be snapped or bolted to a base, which is then mounted on a wall.

Attaching a threaded cap to a lamp base.

FREE-STANDING STAIRS

The landscape will occasionally present situations in which a slope needs to be navigated using stairs. If the slope is steep (more than 3' of fall over 10' of horizontal distance) you should consider building a retaining wall with incorporated stairs. When the slope is gradual (less than the 3' of fall over 10') free-standing stairs can be constructed without the benefit of retaining walls.

For free-standing stairs, large slab like materials can be stacked, overlapping the edge of the higher slab on the lower one. If this design is used, the slab serves as both a riser (the vertical portion of the step) and a tread (the horizontal portion of the step). Each step takes the user a few inches higher up or down the slope. Materials that might be used for this type of step include precast concrete, large flat stones, or other materials with the thickness and large dimensions needed to create safe steps. Materials that have slightly rough surfaces when used as stairs, because they provide better traction, reducing the likelihood of a fall by someone walking up or down them (Figure 9-26).

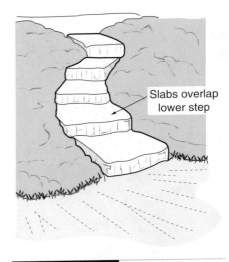

Slabs overlap lower step

Figure 9-26 *Freestanding stairs constructed of slab material.*

Building free-standing stairs with stacked slab material

CAUTION

Select stair material carefully. The thickness of the material will form the risers for the stairs and should be no less than 6" and no more than 9" thick. The flat portion of the material will form the treads and should be no less than 18" front to back to allow

6" of overlap and expose a minimum 12" tread. The entire stair should be at least 24" wide for safety, and a nonslip surface will be the safest.

Time: 2–4 hours for 5 stairs.

Level: Moderate (6 steps). Heavy lifting involved.

Tools Needed:

1. Plan for the project.

2. Marking paint.

3. Round-nosed shovel.

4. Square-nosed shovel.

5. Wheelbarrow.

6. Carpenter's level.

7. Pry bar.

Materials Needed:

1. Large stepping stones, approximately 18" rectangle or 24" diameter round and 6"–9" thick. All material selected should be of a consistent dimension, especially thickness.

2. Angular 1" crushed stone. Approximately 1 CF for each stair planned.

Directions:

1. Use the paint to mark the location for each step on the ground. If the slope is too steep to go up in a straight line, add more steps and lay out a winding or curved alignment.

2. Excavate any sod or ground cover in the location where the steps are to be built.

3. Use the square-nosed shovel to level the area for the steps.

continued

4. Place the lowest stepping stone and level it side to side. The stone should have a very slight slope to the front to allow for water runoff. To adjust the level of the stone, lift it using the pry bar and place a small amount of the angular 1" crushed stone under the low edge. The pry bar can also be used to adjust the placement of the stone (Figure 9-27).

5. Position the second stepping stone with the front edge resting on the back edge of the lower stone. Level the stone side to side, and ensure that it has a slight slope from back to front.

6. Repeat this placement for each stone to the top of the slope.

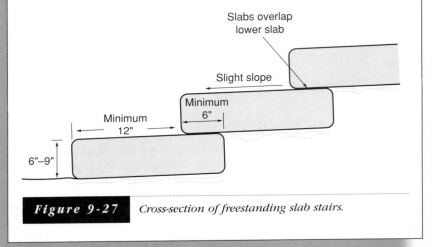

Figure 9-27   *Cross-section of freestanding slab stairs.*

Chapter 10

ADVANCED PROJECTS

As you journeyed through the "projects" in this book, I hope you found new projects to enhance your landscape as well as new ways to enjoy the activities of landscape construction. But don't consider this the conclusion to your "landscaping" experience. Now that you have gained the experience and confidence you may have lacked at the beginning of the book, you are prepared to expand your construction activities. Consider the possibilities suggested in this Advanced Project section. Each idea is intended to add enjoyment and value to your landscape by combining single activities into comprehensive installations.

PROJECTS

To continue the development of your site, consider where in your yard you want to venture next and try some "advanced ideas" that combine projects to develop a complete landscape. Most of these suggestions can be planned and implemented using the same formula presented for the individual tasks.

Begin by looking at projects that address multiple needs with activities from two or more chapters of the book or at multiple projects from the same chapter; for example:

- Try combining a stone wall around a garden that features a dry-laid stone patio. To spice up the combination even more, add an arbor over a portion of the patio.

- Install a kidney-shaped patio that wraps around a pool (be cautious if you have young children to whom the pool may be a risk). To add interest, create a berm to one side and add a cascade that trickles down to the pool (Figure 10-1). A preliminary plan for this installation is included in Appendix D, Figures D-11 and D-12.

■ Place an arbor to the entry of your garden with a stone or brick pathway underneath.

■ Install a brick patio with an adjacent lawn and planting bed. Continue the brick theme using a vertical brick edger around the lawn area.

■ Mix paving materials, using brick to edge a stone patio. Try a vertical brick edge around a granular patio.

■ Select seating and lighting to enhance your garden.

Varying the designs used in this book is also an interesting way to diversify your landscape. Consider modifying the projects provided to create your own special "installations"; some examples are:

■ Rather than putting in a gate, build a 2' × 2' × 3' tall stone pillar (actually just a short wall) on either side of an entry.

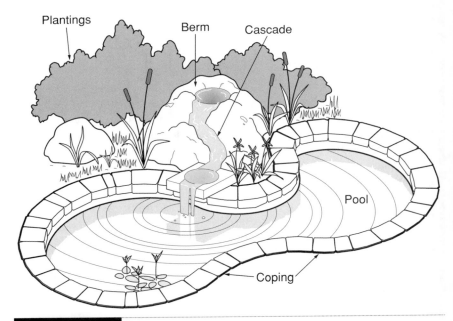

Figure 10-1   *Pool and cascade project. This project combines a flexible-liner pool, a cascade, and a berm project. Plantings and stone are added around the perimeter of the berm and cascade to hide the liners.*

- Recreate the entry to your property with a wood-surfaced stringer fence and an arbor (Figure 10-.2). A preliminary plan for this idea is shown in Appendix D, Figures D-14 and D-15.

- Try more detailed scrollwork or cladding on the posts of your overhead structure. Ideas are suggested in Appendix A, Figures A-5 and A-6.

If you don't like "landscaping" alone, try getting the family and friends involved in landscaping activities. Everyone, including the kids, if you are supervising them, can participate in several of the following activities:

- Let your children work with the paving patterns in a patio.

- Have the kids help sort and grade lumber, pick out amenities, or sweep the joints of your patio.

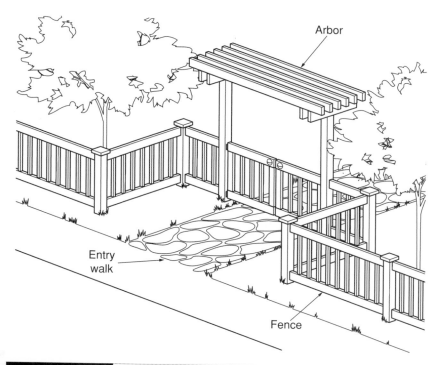

Figure 10-2 *Entry arbor and fence. This project would enhance the entry of many properties. At the gate/arbor, the fence is set back from the walkway a few feet to provide planting space. The arbor accents the entry.*

- Have the entire family work on an edging project.
- Invite friends over for an arbor raising.

If functionality is important to you, the activities described in this text are ideal for addressing the needs of most homeowners; for example:

- Use chainlink fencing to construct a dog kennel or a perimeter fence for the play area.
- Install multi-inlet drainage improvements as solutions to your own water problems. A plan in Appendix D, Figure D-11 shows one method of installing a complex subsurface drainage system.
- Replace an old, broken concrete patio or walk with a safe, attractive, new surface.
- Replace the old perimeter fence with a more attractive wood-panel fence.

FINAL THOUGHTS

As your work concludes, I want to express my sincere hope that you have enjoyed actively participating in the landscaping process. Even the smallest of efforts will help in improving the environment for you, your family, the neighborhood, and for everyone. Whatever purpose you find for completing the projects, keep in mind that the benefits are not just for your landscape. When students who are entering the field ask what they will be paid, my response is always the same. The highest rewards of the profession are obtained not from money but from working with your hands to create something lasting and beautiful.

APPENDIXES

APPENDIX A:
ALTERNATIVE DESIGNS
FOR OVERHEAD ARBORS

Many designs are possible for arbors. In addition to the designs presented in the body of the text, Figures A-1 through A-4 show two alternative designs for overhead arbors. Arbor design 2

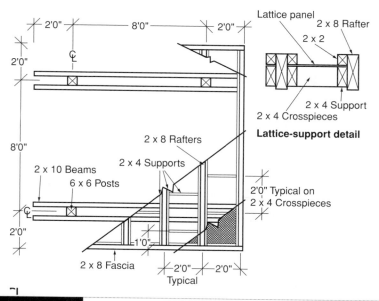

Figure A-1   *Arbor design 2: plan view and lattice-support detail.*

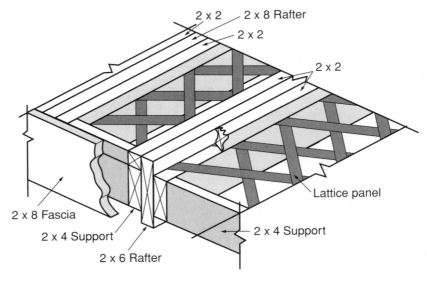

2 x 2 — **2 x 8 Rafter**

2 x 2

2 x 2

2 x 2

Lattice panel

2 x 8 Fascia

2 x 4 Support

2 x 4 Support

2 x 6 Rafter

Perspective View

Figure A-2   *Arbor design 2: elevation and perspective view.*

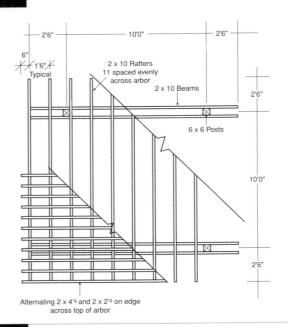

2'6"   10'0"   2'6"

6"
1'6"
Typical

2 x 10 Rafters
11 spaced evenly
across arbor

2 x 10 Beams   2'6"

6 x 6 Posts

10'0"

2'6"

Alternating 2 x 4's and 2 x 2's on edge
across top of arbor

Figure A-3   *Arbor design 3: plan view.*

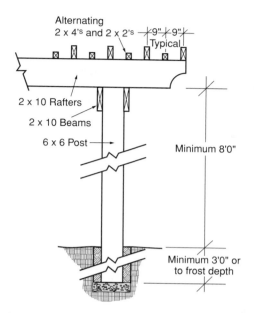

Alternating
2 x 4's and 2 x 2's

9" 9"
Typical

2 x 10 Rafters

2 x 10 Beams

6 x 6 Post

Minimum 8'0"

Minimum 3'0" or
to frost depth

Figure A-4   *Arbor design 3: elevation view.*

depicts a lattice-roofed structure, whereas arbor design 3 is surfaced
with alternating 2 × 4's and 2 × 2's.

To make your standard arbor design more attractive, you can add
accents that convey a higher level of of craftsmanship. Sculpted cuts for
the ends of the beams and rafters of an arbor will provide improved
detail, and cladding the posts for the arbor gives 4 × 4 posts the impres-
sion of being heavier, sturdier supports than they really are (Figures A-5
and A-6).

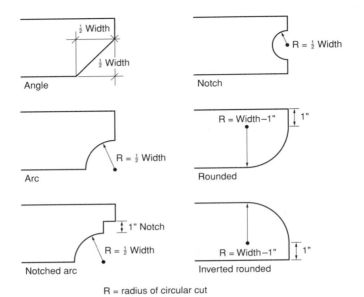

$\frac{1}{2}$ Width

$\frac{1}{2}$ Width

Angle

R = $\frac{1}{2}$ Width

Notch

R = $\frac{1}{2}$ Width

Arc

R = Width−1"   1"

Rounded

1" Notch

R = $\frac{1}{2}$ Width

Notched arc

R = Width−1"   1"

Inverted rounded

R = radius of circular cut

Figure A-5   *End detail for arbor beams and rafters.*

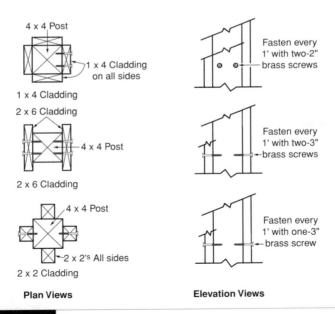

4 x 4 Post

1 x 4 Cladding
on all sides

1 x 4 Cladding

Fasten every
1' with two-2"
brass screws

2 x 6 Cladding

4 x 4 Post

2 x 6 Cladding

Fasten every
1' with two-3"
brass screws

4 x 4 Post

2 x 2's All sides

2 x 2 Cladding

Fasten every
1' with one-3"
brass screw

Plan Views

Elevation Views

Figure A-6   *Post-cladding details.*

APPENDIX B: AN ALTERNATIVE DESIGN FOR A TRELLIS

As with arbors, many designs for trellises are possible. Figures B-7 and B-8 present a contemporary alternative to the wood arbor. The trellis, constructed of clad posts with wire strung between the posts will serve as support for vining plants, as would the wooden trellis. The wire is kept taught using a turnbuckle.

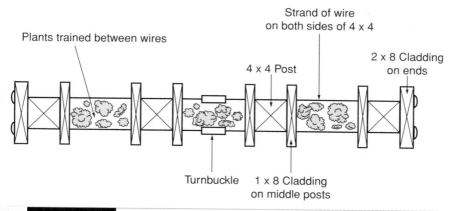

Figure B-7   *Plan for a turnbuckle trellis.*

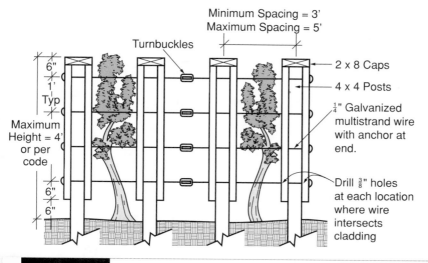

Figure B-8   *Elevation for a turnbuckle trellis.*

APPENDIX C: DESIGNS FOR WOOD-SURFACED FENCES

Few restraints exist when designing the surface of a wood fence. A wide range of horizontal, vertical, and diagonal patterns are possible using common dimensioned lumber. Combining these patterns and creating new arrangements and combinations make possible almost any surface arrangement. Figures C-9 and C-10 provide the construction details for some typical surfacing.

Note: Trim not shown on details

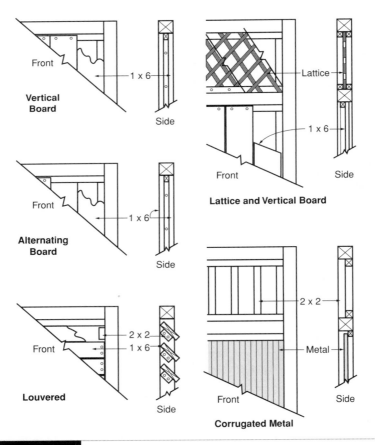

Figure C-9   *Surfaced-stringer fence patterns with stringers hung between posts.*

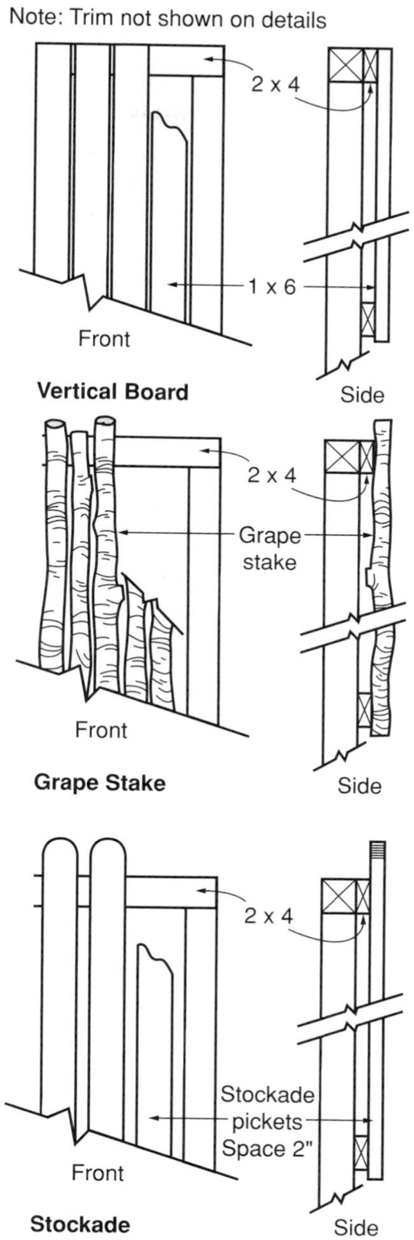

Note: Trim not shown on details

Vertical Board

Front

2 x 4

1 x 6

Side

Grape Stake

Front

2 x 4

Grape stake

Side

Stockade

Front

2 x 4

Stockade pickets
Space 2"

Side

Figure C-10 *Surfaced-stringer fence patterns with stringers fastened to the face of posts.*

APPENDIX D: COMPREHENSIVE LANDSCAPE PROJECTS

To further enhance your landscape, you can combine several techniques into comprehensive projects. Shown in the accompanying illustrations are preliminary designs for three projects that expand or combine projects from separate chapters. Figure D-11 shows an example of how to connect multiple drainage inlets to drain a larger area. Figures D-12 and D-13 show the details for an 8' × 12' pool with a cascade placed on a berm. Figures D-14 and D-15 show the plan and elevation views of a design for an arbor fence used to enhance the entry to a property.

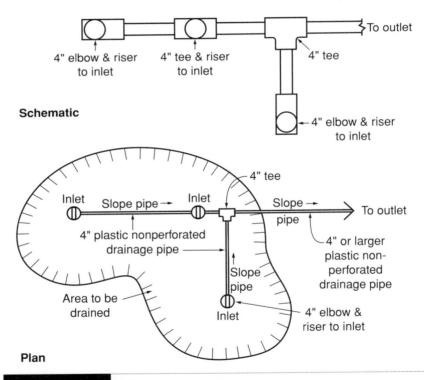

Figure D-11   *Plan of a multi-inlet drainage system.*

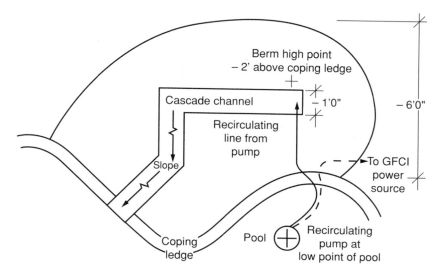

Cascade Detail

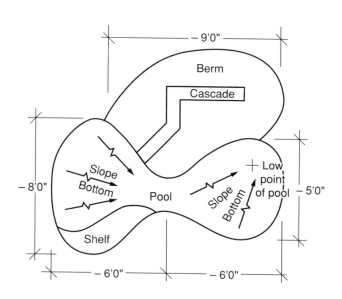

Preliminary Pool Layout

Figure D-12 Pool and cascade project: preliminary pool layout and cascade detail.

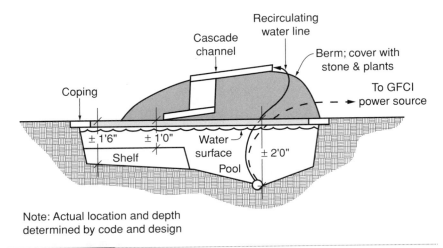

Note: Actual location and depth
determined by code and design

Figure D-13   *Pool and cascade project: conceptual cross-section.*

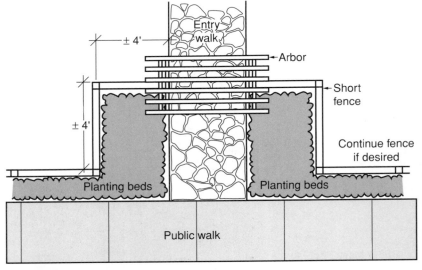

Note: Fence and arbor location and height
determined by code and design

Figure D-14   *Arbor and fence at property entry: plan.*

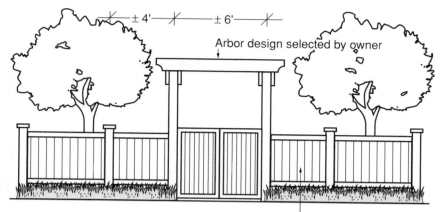

± 4' ± 6'

Arbor design selected by owner

Note: Fence and arbor location and height determined by code and design

Fence design selected by owner

Figure D-15 *Arbor and fence at property entry: elevation.*

GLOSSARY

A

AC (alternating current): Normal household current, which changes direction 60 times per second.

Ag extension: Information source for horticulture questions, typically sponsored by universities.

Amenities: Landscape items such as benches, lights, and bike racks.

Apron: The area surrounding a paved area.

Arbor: An overhead landscape structure built for shade or to support vining plant material.

Architect's scale: A measuring device for plans that uses fractional divisions.

Area calculations: Calculations of surface area using length and width dimensions.

B

Backflow-prevention valve: A valve installed in a water system that prevents water from flowing back toward the supply.

Backhoe: A tractor with a bucket for earthmoving mounted at the rear.

Baseline: An inferred reference line through a project that is used for measuring and locating project elements.

Basemap: A map that locates existing improvements in the project area.

Basket weave: A unit paving pattern that alternates pair of pavers in horizontal and vertical patterns.

Batter: The backward leaning or stepping of a wall.

Berm: A soil mound.

Bonded fiber mulch (BFM): Chemically bonded materials that provide a durable, papier mache-like coating over an exposed area.

Bow saw: An arced saw with large teeth used for pruning and cutting trees.

Brick: Fired-clay rectangular block with typical dimensions of 4" × 8" × 2.5" thick. Available in a range of colors, paving bricks have no mortar holes and an average compression strength of 5000 PSI.

Brick hammer: A hammer with a square head and a long, angled pick that is used for shaping brick.

Brick set: A chisel-like tool with a wide blade

C

Canes: The individual stems of a multistem shrub.

Canopy: The area under the foliage of a tree or shrub.

Capstone: A solid, flat precast concrete block that is placed on top of segmental-unit retaining walls.

Cascade: Water that gently flows on the surface from a higher elevation to a lower one.

Cast-in-place concrete: A mixture of cement, sand, water, and aggregate (stone) that is blended and placed in forms and that hardens to create a solid paving or wall surface.

Chain link: A fencing material that is composed of woven links of medium gauge wire.

Chainsaw: A motor-driven saw with a sharp chainlike rotating blade, normally used to cut trees, branches and large pieces of wood such as landscape timbers.

Cheek wall: A short wall along the sides of stairs.

Chipper: A chamber with gas powered rotating knives that is used to grind wood into small pieces.

Circular saw: An electric saw that cuts using a round blade that spins at high speeds.

Cleave: To cut a brick or stone along a naturally weak line.

Compost: Usable organic matter created by the decomposition of vegetative waste.

Concrete paving block: A unit paving material composed of molded, cured concrete. Pavers come in a variety of shapes (h-shaped, uni-decor, s-shaped) and colors, are 3.5" thick, and have an average compression strength of 8000 PSI.

Crimping: Deforming an object to prevent it falling out of position.

Cross-slope: The slope across a surface.

Composite: A material that is manufactured by blending other materials. Typical landscape composites blend wood and plastics.

Coping: Flat pieces of stone or a similar material used to dress the edges of pools.

Course: Horizontal row of blocks or bricks in a wall.

Cover crops: Fast-germinating plants used to control erosion.

Crushed stone: Stone that has been ground to a smaller dimension than it was originally.

Cutoff saw: A gas-powered saw with a circular blade that is used for heavy-duty cutting.

Cutting: The removal of extra or undesirable soil from a location.

D

DC (direct current): Low-voltage electrical current that flows in only one direction.

Deadman (Deadmen, pl.): Anchors for retaining walls that are placed perpendicular to the face of the wall and anchored into a hillside.

Deck: A surfaced platform, usually constructed of wood, that serves as an outdoor living space; typically one level but can be multilevel.

Dimensioned lumber: Lumber sold by standard width and thickness dimensions, such as 2 × 4's and 4 × 4's.

Drawing scale: A ratio reduction used to depict a large site on a sheet of paper that is convenient to carry and read.

Drainageway: See swale.

Dripline: A line on the ground directly below the outermost foliage of a tree or bush.

Drop spreader: A hand-powered implement that distributes seed and fertilizer by dropping it evenly from the bottom of a bin.

Dry-laid stone: Stone wall pieces that are set without mortared joints.

E

Edge restraint: A material that holds paving in place.

Edging: Strips of material, usually made of plastic or metal, used to hold paving and separate surface materials.

Electrical contractor: A contractor who specializes in electrical work.

Elevations: Project grades, or the vertical heights of landscape elements.

End or corner posts: Heavy-duty posts used to support chain-link fencing; placed at ends and corners of installations.

Engineer's scale: A measuring device for plans that uses multiples of ten.

Erosion: The removal of soil from a site because of the action of water and wind.

Erosion-control blanket (ECB or erosion mats): Fibers or other mulchlike materials sandwiched between layers of thin nylon netting; also called erosion mats.

Erosion Mat: See erosion control-blanket.

F

Felling: The cutting down of a tree.

Field adjustments: Minor changes in a design made when a project is staked.

Fieldstone (rubble): Stone weathered by natural forces.

Filling: Placement of soil where it is needed to bring the surface up to the desired grade.

Fine granular material (termed stone dust or 3/8's minus): Finely ground stone.

Flagstone: Irregularly shaped flat stone used for paving.

Flexible pool liner: A thin sheet of dark, waterproof, impermeable vinyl.

Fountain: A pump with special orifices that project water upward out of a reservoir.

Free-draining angular crushed stone (rubble or washed, class 2 aggregate): Crushed stone or fine granual material that does not have rounded edges used as fill behind a wall.

Free-outs: A low point that allows water to escape from a drainage area.

Free-standing wall: A wall designed to separate a space rather than to retain soil.

French drain: A gravel trench covered with a thin layer of soil and ground cover.

Frost depth: The normal maximum depth of frost in a geographic region.

G

Gabions: Wire baskets that are filled with stone and stacked to create a wall.

Galvanized: Materials, usually metals, that are dipped in a zinc coating to reduce the chance of rusting when exposed to moisture.

Gazebos: Climate-controlled outdoor rooms.

Geogrid: Open-weave fabric used to stabilize retaining walls.

Grading: Shaping soil to a desired form—that is, removing soil from where it is not needed and placing it where it is desired.

Granular backfill: Crushed stone used to fill excavated areas.

Granular paving: Pavement composed of crushed, durable material.

Ground fault circuit interrupter(GFCI): Outlet or circuit that will open (shut off) a circuit if it detects a short.

Gullying: Small valleys caused by erosion.

H

Hardscape: Elements of the landscape that are not living; examples include paving, decks, seating, lighting, and related components.

Hardwoods: Lumber made from from oak, maple, walnut, and other broadleaf trees.

Heartwood: The center portion of a tree that typically produces stronger and more decay-resistant lumber.

Herringbone: A paving pattern composed of units set at perpendicular angles to each other.

Hydraulic stone cutter: A table-mounted masonry cleaving tool that uses hydraulic pressure on a cleaving bar to split materials.

Hydromulching: A process in which mulch is mixed with water and a tackifier (a sticky substance); the resulting mixture is then sprayed on a disturbed area to retard or prevent erosion.

Hydrostatic pressure: The pressure exerted by water.

I

Inlet: A concrete or plastic structure into which water runoff flows.

Item count: The number of each type of item used for a project.

K

Knots: A visual and structural defect of dimensioned lumber created by sawing lumber through the point where a branch connected to a tree.

L

Landscape fabric: Woven fabric used to control weeds in the landscape.

Landscape plan (site plan): A drawing that sketches out the improvements for a project.

Land surveyor: A registered professional who surveys land for grades and dimensions.

Lattice: A gridded panel, typically made of small strips of wood or vinyl.

Lean-back batter: The backward lean of a wall in order to aid in stabilizing the it. Accomplished by leaning the base course slightly backward which causes all courses above to lean the same direction.

Linear measurements: Quantities for items that are purchased by length.

Line posts: Lightweight posts used in chain-link-fence installations; placed in line with end or corner posts.

Lipped blocks: Precast wall materials that include a protrusion on the bottom of the block that helps to integrate layers.

Loading boom: A hydraulic arm mounted on a truck and used to load and unload heavy objects.

Loading equipment: Vehicles equipped with forks to lift and move landscaping materials.

Lopping shears: A hand-operated cutting tool used to cut smaller branches.

M

Magnesium float: A wide, flat tool with a handle mounted on top used to smooth concrete surfaces; made of lightweight magnesium.

Masonry: Paving or wall materials secured with mortar.

Masonry bit: A hardened bit used to drill into concrete and masonry materials.

Material take-off: Calculation of the quantities of all materials to be used for a project.

Mortared veneer: Concrete walls with stone or brick mortared on the front face to make them more attractive.

Mulching: Organic or inorganic ground covering placed around plants to retain moisture and discourage the growth of weeds .

N

Niches: Offsets or corners in a fence.

Nonperforated plastic pipe: Solid plastic drainage pipe.

O

Ordinances: Regulations that place stipulations or limitations on the use of land and the construction of improvements on property.

Orifices: Openings.

P

Paper form: A heavy-duty cylindrical paper tube used to form concrete.

Perforated plastic pipe: Plastic drainage pipe with holes along the sides that allow water to enter.

Perimeter: The distance around an object or space.

Photocell: An electrical control device that turns circuits on and off based on the presence of light.

Pilot holes: Holes drilled into, or through, wood pieces that ease the installation or screws or bolts.

Plan view: A drawing depicting a project as if looking straight down at it from above.

Pneumatic jackhammer: An air-powered vertical hammer used to break up solid surfaces.

Polymer-coated fasteners: Materials that have been coated with a thin plasticlike coating formulated to resist rust.

Pool: Water reservoirs with little or no moving water.

Post-hole excavator (auger or clamshell): Tools, powered or hand-operated, used to excavate vertical holes for installing posts; augers work like drills to remove soil, whereas clamshells work like pincers to cut and remove soil.

Pruning: Selective removal of portions of a plant.

Pruning saw: A curved hand saw used for cutting branches.

R

Rail: A horizontal board, or component, of a fence.

Rammer plate: A powered implement that uses vertical pounding to compact materials.

Random irregular: Paving pattern for stone that fits together irregular pieces of stone.

Reinforcing materials: Metals and fiberglass materials that are placed in concrete to add strength.

Reinforcing rod (rerod or rebar): Molded steel bars with ribbing that holds them in place in reinforced concrete pours. Rods come in lengths up to 20' and diameters from 3/8" (#3) to over 1".

Retaining wall: A wall used to hold back earth.

Retreatment: Painting or dipping the cut portions of a treated board.

Rip-rap: Large stone used to stabilize erodable surfaces.

Riser: The vertical portion of a step.

Rototiller: A powered, walk-behind implement with rotating blades used to break up soil.

Rough-sawn: Wood that has been cut but not planed to a smooth surface; the rough surface is left for esthetic effect.

Running bond: A wall and paving pattern in which each subsequent course is offset by half from the previous course.

S

Screeding: The horizontal leveling of a surface using a straight bar or piece of wood.

Segmental wall blocks (pre-cast concrete): Pre-cast concrete blocks used for wall construction.

Setting bed: A layer of material between the base and the paver which helps to keep the paved surface smooth and level.

Shanks: The long, rounded portion of a nail, screw, or bolt.

Shelf: A flat ledge constructed in a pool to create multiple levels.

Short: An electrical flaw in which a circuit is interrupted by an electricity-conducting material (typically a wire or nail) touching it, creating an alternative path for the flow of electrical current.

Skid steer: A four-wheeled or tracked tractor with a bucket; steering is accomplished by braking wheels on one side.

Socked pipe: A drainage pipe enclosed in a fabric coating to prevent soil and debris from entering it.

Sod cutter: A powered, walkbehind implement with a single oscillating blade that undercuts turf to make it easier to remove.

Sod staples: U-shaped metal staples used to anchor sod by driving them through sod into the ground.

Softscape: Plant material used in landscaping.

Softwoods: A term used to refer to lumber harvested from conifer trees such as pine, fir, and spruce

Soldier course: A perimeter paving course that is set perpendicular to the pavement edge; used to stabilize the cut edges of paving blocks.

Spikes: Long nails.

Stacked bond: A paving or wall pattern in which units are stacked one on top of another.

Step-back batter: An anchoring method for walls in which the front of each subsequent course is set back a set distance from the course beneath it.

Stepping stones: Loose pieces of paving arranged in a convenient stepping pattern.

Stretcher bar: An thin, 1/2" wide metal bar inserted vertically into chain-link fence fabric to aid in connection to an end post.

Stretchers: A wide, jacklike tool used to stretch or tighten chain-link fence fabric.

Stringline: String set up across a site to check elevations.

Structural members (lumber): Large-dimensioned pieces of lumber that support roofs, fences, and decks.

Subgrade: The soil below the base of a wall or paving project.

Submersible fountain: A fountain that operates under water.

Surface pattern (staggered, or running-bond placement): A wall pattern in which each unit is offset by half from the previous course.

Swale: A shallow drainage ditch.

T

Tackifier: A sticky oil based substance used to hold bonded fiber mulches together.

Tapering: The gradual reduction of grade.

Tension wire: A wire connected to the bottom of chain-link fabric to prevent curling.

Terraces: A series of retaining walls.

3,4,5 triangle: A method used for locating an object at a right angle to a line by constructing a triangle with sides in the ratio of 3:4:5.

Top rail: The top piping of a chain-link fence used to support the top of the fabric.

Transformer: An electrical device that changes high–voltage AC current to low-voltage DC current.

Tread: The horizontal portion of a step.

Triangulation: The process of locating an object by measuring from two known points.

Treated: Wood that has been dipped or injected with chemical compounds to resist decay and insect attack.

Trellis: A vertical or wall-like landscape structure, typically built from wood and used as a framework on which vining plants can grow.

Topsoil: The upper layer of soil, typically with a high organic-matter content.

U

Unit paver: A collective term that includes clay bricks, interlocking concrete paving blocks, adobe pavers, precast concrete units, and similar types of materials that are produced and installed as individual pieces.

V

Vellum: Translucent drawing paper.

Vibratory plate compactor (vibraplate): A metal plate with a motor mounted on top and a handle for steering it; vibrates and compacts base material as the machine moves.

Volume measurement: The bulk measurement of material.

W

Washed river rock: Stone that has been smoothed and rounded by being tumbled in a river.

Water table: Subsurface water that is moving up toward the surface.

Weep holes: Holes drilled through a wall to allow water to drain from behind the wall.

Wet-masonry saw: A table-mounted circular saw that uses water to cool the blade as it cuts masonry materials.

Wood landscape timber: A dimensioned lumber wall unit treated for contact with ground to resist decay and insect damage.

Wood-panel fencing: Wood fencing composed of segments of wood surfacing supported by posts.

INDEX